Preparing the Next Generation of Earth Scientists

AN EXAMINATION OF FEDERAL EDUCATION AND TRAINING PROGRAMS

Committee on Trends and Opportunities in Federal Earth Science Education and Workforce Development

Board on Earth Sciences and Resources

Division on Earth and Life Studies

NATIONAL RESEARCH COUNCIL
OF THE NATIONAL ACADEMIES

THE NATIONAL ACADEMIES PRESS
Washington, D.C.
www.nap.edu

THE NATIONAL ACADEMIES PRESS 500 Fifth Street, NW Washington, DC 20001

NOTICE: The project that is the subject of this report was approved by the Governing Board of the National Research Council, whose members are drawn from the councils of the National Academy of Sciences, the National Academy of Engineering, and the Institute of Medicine. The members of the committee responsible for the report were chosen for their special competences and with regard for appropriate balance.

The project described in this publication was supported by Grant/Cooperative Agreement Number G11AP20216 from the U.S. Geological Survey. Its contents are solely the responsibility of the authors and do not necessarily represent the official views of the USGS.

International Standard Book Number-13: 978-0-309-28747-0
International Standard Book Number-10: 0-309-28747-2

Cover: Photo strip: Students learning about geology in the field. (Left) Mammoth Cave National Park, National Association of Geoscience Teachers (nagt.org/details/images/26083.html); (Center and Right) Cedar Breaks National Monument, National Park Service. Background image: Copyright by Michael Dudzik.

Additional copies of this report are available for sale from the National Academies Press, 500 Fifth Street, NW, Keck 360, Washington, DC 20001; (800) 624-6242 or (202) 334-3313; http://www.nap.edu/.

Printed in the United States of America.

THE NATIONAL ACADEMIES

Advisers to the Nation on Science, Engineering, and Medicine

The **National Academy of Sciences** is a private, nonprofit, self-perpetuating society of distinguished scholars engaged in scientific and engineering research, dedicated to the furtherance of science and technology and to their use for the general welfare. Upon the authority of the charter granted to it by the Congress in 1863, the Academy has a mandate that requires it to advise the federal government on scientific and technical matters. Dr. Ralph J. Cicerone is president of the National Academy of Sciences.

The **National Academy of Engineering** was established in 1964, under the charter of the National Academy of Sciences, as a parallel organization of outstanding engineers. It is autonomous in its administration and in the selection of its members, sharing with the National Academy of Sciences the responsibility for advising the federal government. The National Academy of Engineering also sponsors engineering programs aimed at meeting national needs, encourages education and research, and recognizes the superior achievements of engineers. Dr. C. D. Mote, Jr., is president of the National Academy of Engineering.

The **Institute of Medicine** was established in 1970 by the National Academy of Sciences to secure the services of eminent members of appropriate professions in the examination of policy matters pertaining to the health of the public. The Institute acts under the responsibility given to the National Academy of Sciences by its congressional charter to be an adviser to the federal government and, upon its own initiative, to identify issues of medical care, research, and education. Dr. Harvey V. Fineberg is president of the Institute of Medicine.

The **National Research Council** was organized by the National Academy of Sciences in 1916 to associate the broad community of science and technology with the Academy's purposes of furthering knowledge and advising the federal government. Functioning in accordance with general policies determined by the Academy, the Council has become the principal operating agency of both the National Academy of Sciences and the National Academy of Engineering in providing services to the government, the public, and the scientific and engineering communities. The Council is administered jointly by both Academies and the Institute of Medicine. Dr. Ralph J. Cicerone and Dr. C. D. Mote, Jr., are chair and vice chair, respectively, of the National Research Council.

www.national-academies.org

COMMITTEE ON TRENDS AND OPPORTUNITIES IN FEDERAL EARTH SCIENCE EDUCATION AND WORKFORCE DEVELOPMENT

ARTHUR GOLDSTEIN, *Chair*, Bridgewater State University, Bridgewater, Massachusetts
PRANOTI ASHER, American Geophysical Union, Washington, D.C.
SUSAN E. COZZENS, Georgia Institute of Technology, Atlanta
CATHRYN A. MANDUCA, Carleton College, Northfield, Minnesota
ERIC J. PYLE, James Madison University, Harrisonburg, Virginia
ERIC M. RIGGS, Texas A&M University, College Station
KARL TUREKIAN, emeritus, Yale University, New Haven, Connecticut (deceased March 2013)
LISA D. WHITE, University of California Museum of Paleontology, Berkeley

National Research Council Staff

ANNE M. LINN, Study Director, Board on Earth Sciences and Resources
DEBORAH GLICKSON, Senior Program Officer, Ocean Studies Board
ERIC J. EDKIN, Senior Program Assistant, Board on Earth Sciences and Resources

BOARD ON EARTH SCIENCES AND RESOURCES

Acknowledgments

This report has been reviewed in draft form by individuals chosen for their diverse perspectives and technical expertise, in accordance with procedures approved by the NRC's Report Review Committee. The purpose of this independent review is to provide candid and critical comments that will assist the institution in making its published report as sound as possible and to ensure that the report meets institutional standards for objectivity, evidence, and responsiveness to the study charge. The review comments and draft manuscript remain confidential to protect the integrity of the deliberative process. We wish to thank the following individuals for their participation in the review of this report:

Matthew Dawson, The Geological Society of America, Boulder, Colorado
Elizabeth Day-Miller, BridgeWater Education Consulting, LLC, Bridgewater, Virginia
Lisa Grant Ludwig, University of California, Irvine
Mitchell Leverette, Bureau of Land Management, Washington, D.C.
Roger Levine, American Institutes of Research, retired, Redwood City, California
R. Heather MacDonald, College of William and Mary, Williamsburg, Virginia
Camille McKayle, University of the Virgin Islands, St. Thomas, U.S. Virgin Islands

Although the reviewers listed above have provided many constructive comments and suggestions, they were not asked to endorse the conclusions or recommendations nor did they see the final draft of the report before its release. The review of this report was overseen by George M. Hornberger, Vanderbilt University, and Katherine H. Freeman, Pennsylvania State University. Appointed by the National Research Council, they were responsible for making certain that an independent examination of this report was carried out in accordance with institutional procedures and that all review comments were carefully considered. Responsibility for the final content of this report rests entirely with the authoring committee and the institution.

Contents

APPENDIXES

Summary

Earth science (defined here as excluding oceanic, atmospheric, and space science) plays a key role in the well-being of our nation, and many issues in its purview—including hydrocarbon, mineral, and water resources; the environment; and geological hazards such as earthquakes and volcanic eruptions—are expected to grow in importance. Addressing these issues requires an earth science workforce that draws on the talents of all citizens, including women and minorities historically underrepresented in earth science. Federal education programs can help attract students to or retain them on an earth science pathway, but with tight funding, it is imperative for agencies to invest in programs that work. At the request of the U.S. Geological Survey (USGS) Office of Science Quality and Integrity, the National Research Council (NRC) established a committee to carry out a study, organized around a workshop, to address the following tasks:

1. Summarize the legislative authority for science, technology, engineering, and mathematics (STEM) education and training granted to federal agencies with substantial programs in earth science (excluding oceanic, atmospheric, and space science).
2. Examine recent earth science education programs with a research or training component, both formal and informal, in these federal agencies.
3. Identify criteria for evaluating the success of earth science education and training programs and, using these criteria and the results of previous federal program evaluations, identify examples of successful programs in federal agencies.
4. Determine what made these example programs successful (e.g., resources, themes, engagement activities, partnerships).
5. Summarize the knowledge and skills identified in recent NRC workforce reports that are needed by earth scientists in their careers.
6. Describe ways that federal agencies can leverage their earth science education and training efforts to improve their recruitment of a diverse population in both high school and college.

Information for these tasks was provided by federal agency managers and drawn from published articles and reports. Federal earth science education programs and efforts to leverage resources were examined at a 2-day workshop attended by government, academic, and professional society managers of earth science education and outreach programs, and experts knowledgeable about education, the transition to earth science careers, and program evaluation. The committee's main conclusions about the six tasks are summarized below.

FEDERAL EARTH SCIENCE EDUCATION PROGRAMS

Legislative Authority for STEM Education

The first task of the committee was to summarize the legislative authorities for STEM education granted to federal agencies with substantial programs in earth science. These agencies include the USGS, National Science Foundation (NSF), Department of Energy (DOE), National Aeronautics and Space Administration (NASA), U.S. Department of Agriculture (USDA), Environmental Protection Agency (EPA), National Oceanic and Atmospheric Administration (NOAA), National Park Service, and the Smithsonian Institution. All of these agencies have legislative authority for STEM education, although the scope of authority varies widely. For example, NASA and NOAA are authorized to support research-based programs to increase student participation in STEM, and NSF is authorized to strengthen education in all science areas covered by the foundation. A few agencies have specific authorization for earth science education programs. Examples include authorization for the USGS to support education in geologic mapping and field analysis, for DOE to promote education and training in methane hydrate resources, and for NASA to fund museum and planetarium programs related to earth science and other fields. Other earth science education programs are created to support agency missions, to help build a pool of potential recruits, or to fulfill other agency objectives.

Federal Earth Science Education Programs with a Research or Training Component

For Task 2, the committee examined 25 federal earth science education programs identified by their host agencies as having a research or training component (Box S.1). Given the research or training criterion, most agencies identified education programs aimed primarily at high school and college students. Because of time and budget constraints, the committee neither considered other programs that might fit these criteria nor culled the agency-identified programs. Although the set of programs considered in this report is not comprehensive, it covers a wide range of objectives and audiences and led the committee to develop a conceptual framework for thinking about all earth science education programs.

The earth science pathway can be thought of as a system of opportunities and experiences that attract individuals to the field and prepare them for employment. This conceptual framework is illustrated in Figure S.1. In this framework, individuals become aware of earth science, then engage in learning about the Earth and the nature of earth science, and finally prepare for a career by acquiring specialized knowledge, skills, and expertise and by exploring different employment options. The various education and training opportunities are represented by the upward-pointing polygons in Figure S.1. Some programs have multiple objectives and span more than one stage of the framework. The framework is portrayed as a triangle because more individuals will develop an interest in earth science than will become engaged in the discipline, and more will become engaged in the field than will prepare for a career.

The federal earth science education programs identified by the agencies are situated at every stage of the framework. At the awareness stage, formal education and informal learning through museums and after-school programs and clubs bring earth science to the attention of individuals and spark their interest. Examples of such activities include the USDA's 4-H club, which works

BOX S.1 Federal Earth Science Education and Training Programs Considered in This Report

U.S. Geological Survey

- National Cooperative Geologic Mapping Program (EdMap)
- Cooperative Summer Field Training Program (with the National Association of Geoscience Teachers [NAGT])
- Youth Internship Program
- Hydrologic Technician Internship Program

National Science Foundation

- Earth Science Research Experience for Undergraduates (REU) Program
- Opportunities for Enhancing Diversity in the Geosciences (OEDG) Program
- Earth Sciences Postdoctoral Fellowships
- Geoscience Education (GeoEd) Program
- Global Learning and Observations to Benefit the Environment (GLOBE) Program (with NOAA and NASA)
- Geoscience Teacher Training (GEO-Teach) Program

Department of Energy

- Office of Science Graduate Fellowship (SCGF) Program
- Summer of Applied Geophysical Experience (SAGE) Program
- Science Undergraduate Laboratory Internships (SULI) Program
- Community College Internships (CCI) Program

National Aeronautics and Space Administration

- Gravity Recovery and Climate Experiment (GRACE) programs

U.S. Department of Agriculture

- 4-H Environmental Education/Earth Science programs
- Agriculture and Food Research Initiative (AFRI) projects
- AFRI National Institute of Food and Agriculture Fellowships Grant Program

Environmental Protection Agency

- Greater Research Opportunities (GRO) Undergraduate Fellowship Program
- Science to Achieve Results (STAR) Graduate Fellowship Program

National Oceanic and Atmospheric Administration

- Educational Partnership Program (EPP) with Minority-Serving Institutions

National Park Service

- Geoscientists-in-the-Parks Program (with the Geological Society of America)
- Geoscience-Teachers-in-Parks Program (with NAGT)
- National Fossil Day

Smithsonian Institution

- Leadership and Assistance for Science Education Reform (LASER) Program

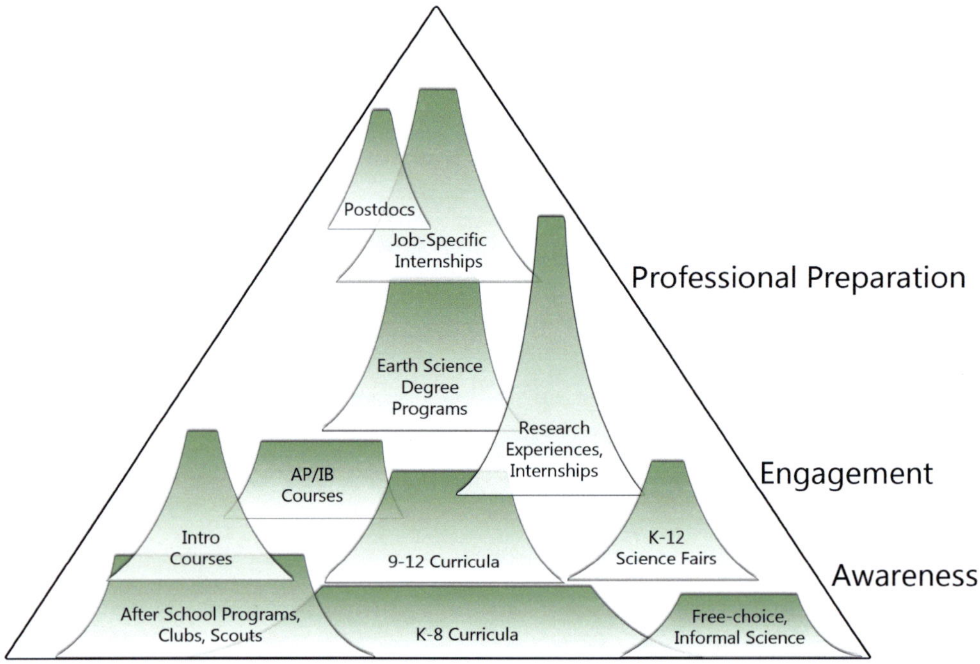

FIGURE S.1 The committee's conceptual framework illustrating the types of education opportunities and experiences (tapering polygons) that individuals encounter along a path from awareness of earth science (base of the triangle) toward an earth science career (apex of the triangle). Polygons are not drawn to scale, but their vertical extent is intended to show that some education opportunities span more than one stage of the framework and their relative horizontal extent is intended to show that more individuals participate in awareness activities than in professional preparation activities.

to increase science awareness among youth, and the National Park Service's National Fossil Day, which promotes public awareness of the scientific and educational value of fossils.

At the engagement stage, students actively engage in learning earth science by choosing earth science-related courses, research, clubs, or community service activities. Federal programs with engagement objectives include the USGS Youth Internship Program, which offers hands-on earth science projects, and NSF's OEDG Program, which supports projects aimed at expanding the interest of underrepresented groups in earth science.

The transition from engagement to professional preparation occurs when an individual shifts focus from exploring earth science to acquiring job skills, knowledge, and abilities. Federally sponsored research experiences help students build skills and expertise in a specialty area (e.g., NSF's REU Program, EPA's GRO program, DOE's SULI Program). Internships (e.g., USGS Hydrologic Technician Internship Program, DOE's CCI Program) and postdoctoral positions introduce students and early-career scientists to job opportunities and employers and provide work-related skills.

Although portrayed as a linear progression, the path through the framework may be full of twists and detours, and individuals may enter or leave the path at different points and for different reasons. Research suggests that specific events, commonly called critical incidents, can lead individuals into certain educational and career paths. For example, some individuals discover an interest in earth science before they reach college, commonly through activities such as family trips to geologically interesting areas. Others discover earth science as an attractive field of study in college or later, commonly through an outstanding introductory course. Understanding these different populations and pathways could help federal agencies design awareness and engagement programs that attract and retain a wide range of individuals in earth science.

Students are responsible for finding education opportunities that move them along a path to the workforce. Federal agencies could help students navigate a path toward an earth science career by improving program visibility and person-to-person connections among their programs. Networks that link people and programs are especially important for attracting and retaining students from underrepresented groups. A connected system of federal, academic, and professional society programs would increase the visibility of a variety of available earth science education opportunities as well as help students move from interest to employment.

PROGRAM EVALUATION

Identifying Successful Programs

The committee's third task was to develop criteria for success and then to use the criteria and the results of previous program evaluations to identify successful examples of federal earth science education programs. Because criteria for success depend on program goals, which range from raising awareness to professional preparation, no single set of criteria can be used to determine the success of all education programs considered in this report. Rather, demonstrating program success requires a comprehensive evaluation approach aimed at understanding program goals, establishing criteria for success, and gathering data to compare program performance to the criteria for success. This approach has been used to demonstrate the success of the OEDG Program, the effectiveness of the selection process for STAR fellows, and progress toward achieving EPP goals. The other federal programs considered in this report have not been formally evaluated and most were not designed to facilitate evaluation. For example, some program goals are too broad to develop criteria for success; the goals and criteria do not always match; and the criteria and data collection emphasize what is easy to measure, not what the program is trying to achieve. Although these programs may be successful, the data were too sparse and uneven in quality for the committee to make an independent determination.

The lack of suitable data for identifying successful programs underscores the importance of incorporating evaluation into program design. Logic models provide a useful mechanism for helping program managers define who the program is trying to reach, what it is trying to achieve, what resources it requires (inputs), and how to translate program resources into near-term results (outputs) and long-term outcomes. Examples of input, output, and outcome measures for programs at the awareness, engagement, and professional preparation stages of the framework are given in the body of the report, although the most useful measures will depend on each program's particular goals. Periodic evaluations would help managers determine whether activities are aligned with program goals and the extent to which those goals are being met. Enumeration (counting participants or their characteristics), pre- and post-testing, observations of participants or providers, work product analysis, and determination of long-term plans and satisfaction with experiences are all useful tools for evaluation.

A system-level evaluation, encompassing all activities within the framework or at a stage of the framework (e.g., engagement), could be used to identify imbalances in effort and gaps, enabling agencies to determine where future education and training efforts may be useful. In such assessments, information from individual program evaluations could be aggregated and supplemented with targeted program evaluations aimed at understanding how to create effective programs. Network analysis of the programs in the system could reveal which connections among participating organizations help move individuals through the system, and qualitative studies would help show how individuals find education and training opportunities and what they learn from them.

Why Programs Are Successful

Given the limited information available to identify examples of successful federal earth science education programs, the committee could offer only limited insight on why these programs might be successful (Task 4). The most common factors identified by federal program managers were stable funding, cost sharing, the commitment of agency managers or principal investigators, and partnerships. Agency support, community outreach, and program design (e.g., a good fit between participants and providers, flexibility, institutionalization) were viewed as important for the success of some programs. All of these factors are reasonable and consistent with workshop discussions and the literature.

KNOWLEDGE AND SKILLS NEEDED FOR EARTH SCIENCE CAREERS

Task 5 was to summarize the knowledge and skills needed by earth scientists in their careers, as identified in recent NRC workforce reports. Only two NRC reports examine the earth science workforce explicitly and they only touch on the knowledge and skills required for a few subdisciplines. *Emerging Workforce Trends in the U.S. Energy and Mining Industries: A Call to Action* (NRC, 2013a) examines the oil and gas, mining, and geothermal energy industries. The report does not discuss specific knowledge or skills for these industries, but stresses the importance of a strong foundation in STEM—including applied mathematics, reading for information, and locating information. *Future U.S. Workforce for Geospatial Intelligence* (NRC, 2013b) contains a discussion of geodesy and geophysics in the context of national intelligence. Important knowledge and skills for geodesy include mathematics, the principles of gravity field theory and orbital mechanics, the propagation of electromagnetic waves, and the theory and operation of observing instruments such as Global Navigation Satellite System receivers. For geophysics, important knowledge and skills include mathematics and the principles of physics, geodesy, seismology, the structure and evolution of the Earth, the theory and measurement of the Earth's magnetic field, and space physics. Both fields require basic interpersonal skills, effective communication, and creative thinking.

BROADENING PARTICIPATION OF UNDERREPRESENTED GROUPS

Task 6 of the committee was to describe ways federal agencies can leverage their earth science education and training efforts to improve their recruitment of a diverse population in both high school and college. A number of federal programs are aimed at increasing the ethnic, racial, and gender diversity of earth science pathways. Women, who make up 51 percent of the U.S. population, have made substantial gains in earth science over the past several decades and now receive approximately 40 percent of bachelor's degrees. With attention to mentoring and unconscious biases, it may be possible to further narrow or eliminate the degree gap between women and men.

Increases in the number of minorities receiving earth sciences degrees have been modest. Underrepresented minorities (African American, American Indian, and Hispanic or Latino of any race) compose 30 percent of the U.S. population, but receive only about 7 percent of earth science bachelor's degrees. Programs that raise awareness of earth science or that increase access to education and training (e.g., social and professional networks, financial assistance for study) may be especially fruitful for federal agencies looking to increase diversity. The importance of connections among programs and between programs and communities underscores the utility of thinking about federal earth science education and training programs in the context of a system of opportunities that creates clear educational pathways for students.

A system approach would also help federal agencies leverage resources. By mapping their diversity programs onto a common conceptual framework, such as the one illustrated in Figure S.1, agencies could identify potential partners and share effective practices for attracting and retaining minority students. Collaborations with professional societies focused on diversity (e.g., National Association of Black Geoscientists) could help connect minority students to education and training opportunities, providing students with another avenue of information on available positions. Coalitions of partners from federal agencies, private companies, universities, and professional societies would stretch federal dollars and bring a wide range of expertise to bear on training the next generation of earth scientists. Although it takes time to build trust and establish common goals and approaches, such partnerships could both benefit the profession and help federal agencies meet their missions.

MAJOR CONCLUSIONS

The 25 federal earth science education and training programs considered in this report provide a wide range of opportunities that interest students and citizens in earth science, engage students in study of the Earth, and prepare them for earth science careers. Such efforts also contribute toward meeting national goals of developing a robust, diverse earth science workforce. Widening earth science pathways requires both a variety of programs that work and connections between programs that help move students through formal and informal education to the workforce. Although some earth science education programs considered in this report have demonstrated success through rigorous evaluation approaches, others were not designed to facilitate evaluation and have not collected the data needed to determine whether they have succeeded and why. In addition, connections between programs are not yet robust.

A conceptual framework, such as the one illustrated in Figure S.1, shows how the various education opportunities fit together and where connections are needed to move students along earth science pathways. Such connections are particularly important for increasing diversity, but benefit all students as well as earth science employers. Other benefits of a conceptual framework include helping federal agencies determine where to focus effort by identifying gaps, overlaps, and imbalances of effort among programs; where to find potential partners in other agencies or organizations to share work and resources; and supporting evaluation of programs at the various stages of the framework and of the system of opportunities.

1

Introduction

Earth science (defined here as excluding oceanic, atmospheric, and space science) plays a key role in the well-being of our nation, and many issues in its purview—including resources, the environment, and geological hazards—are expected to grow in importance in the future. Our needs for hydrocarbon, mineral, and water resources are increasing. As we turn toward nontraditional sources of hydrocarbons, such as shale gas and deep offshore oil reservoirs, and seek the metals and minerals needed to build modern electronic devices, the United States will need earth scientists not only to discover and exploit those resources but also to monitor the environmental consequences of their extraction. Similarly, earth scientists will be needed to monitor the availability and quality of water for drinking, irrigation, and industrial uses. Water is already scarce in some regions of the country, and the drought of 2012 brought into focus the sensitivity of our food supply to changing environmental conditions. Severe droughts may become more common as the climate changes (IPCC, 2012), and the geologic record provides insight on the history and extent of drought. Finally, growing numbers of people are living in geologically hazardous areas, increasing the importance of providing scientific information to help affected populations prepare for earthquakes and tsunamis, severe coastal storms, landslides, and volcanic eruptions.

Addressing these and other earth science issues requires a well-educated and -trained workforce. The Bureau of Labor Statistics projects that job growth will increase by 21 percent for geoscientists (geologists and geophysicists) and by 18 percent for hydrologists from 2010 to 2020, compared with 14 percent for all occupations.[1] Despite high projected demand for earth scientists, however, the number of graduates in earth science fields has not fully recovered from a sharp decline in the early 1980s, which was caused by a loss of U.S. jobs in the petroleum and mining industries (Figure 1.1).

A robust workforce also harnesses the talents of all citizens. Although the fraction of women earning bachelor's degrees in geoscience (earth science plus environmental, ocean, atmospheric, and climate science) has grown to 39 percent, the fraction of underrepresented minority (Black, Hispanic, American Indian/Alaskan Native) graduates remains about 7 percent (Gonzales and Keane,

[1] See http://bls.gov/news.release/ecopro.toc.htm.

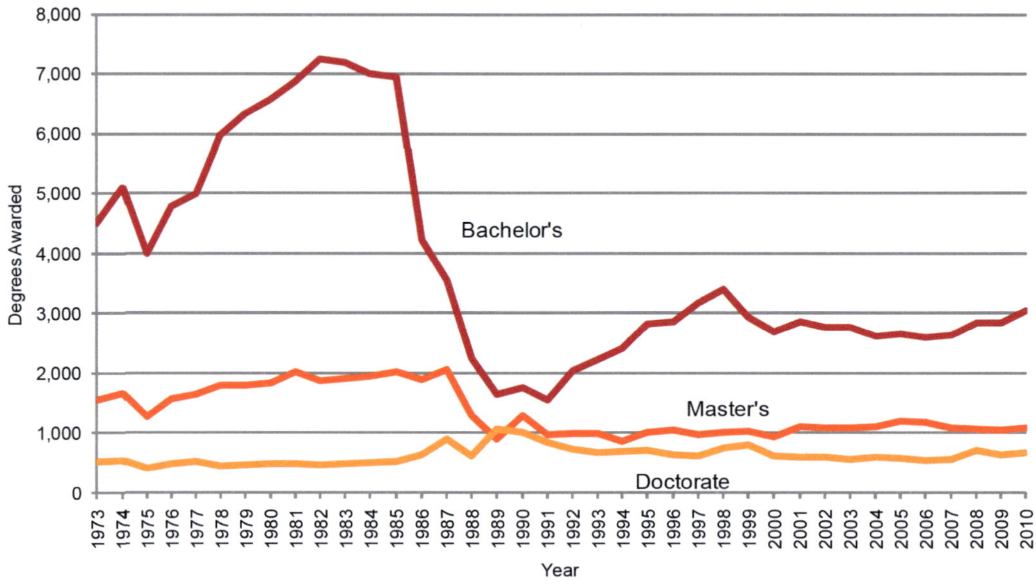

FIGURE 1.1 Trends in the number of geoscience degrees (defined in this figure as encompassing environmental science, hydrology, oceanography, atmospheric science, geology, geophysics, climate science, geochemistry, paleontology; environmental, exploration, and technical engineering; and geoscience management) awarded at U.S. 4-year colleges from 1973 to 2009. SOURCE: Gonzales and Keane (2011).

2011). Neither population is well represented in the geoscience workforce. In 2009, women held 30 percent of environmental science and geoscience jobs and underrepresented minorities held less than 8 percent (Gonzales and Keane, 2011).

To help increase the number, quality, and diversity of earth science graduates, federal agencies that hire earth scientists are investing in a variety of education and training programs. Education funding is commonly scarce, so it is imperative that these efforts focus on programs that work. At the request of the U.S. Geological Survey (USGS) Office of Science Quality and Integrity, the National Research Council (NRC) established a committee to carry out a study, organized around a workshop, to address the following tasks:

1. Summarize the legislative authority for science, technology, engineering, and mathematics (STEM) education and training granted to federal agencies with substantial programs in earth science (excluding oceanic, atmospheric, and space science).
2. Examine recent earth science education programs with a research or training component, both formal and informal, in these federal agencies.
3. Identify criteria for evaluating the success of earth science education and training programs and, using these criteria and the results of previous federal program evaluations, identify examples of successful programs in federal agencies.
4. Determine what made these example programs successful (e.g., resources, themes, engagement activities, partnerships).
5. Summarize the knowledge and skills identified in recent NRC workforce reports that are needed by earth scientists in their careers.
6. Describe ways that federal agencies can leverage their earth science education and training efforts to improve their recruitment of a diverse population in both high school and college.

COMMITTEE APPROACH

The committee began its work by compiling the legislative authorities granted to federal earth science agencies for STEM education (Task 1 of the committee charge) published in reports (e.g., Co-STEM, 2012) and agency planning documents. Federal agencies with substantial earth science programs include the USGS, National Science Foundation, Department of Energy, National Aeronautics and Space Administration, U.S. Department of Agriculture, Environmental Protection Agency, National Oceanic and Atmospheric Administration, National Park Service, and the Smithsonian Institution. For Task 2, the committee asked these federal agencies to identify earth science education programs that have a research or training component (e.g., by providing research experiences to students). Education programs tied to research or training are commonly aimed at high school and college students, although a few agencies identified programs aimed at elementary and middle school students. Given the time and budgetary constraints for the study, the committee neither considered other programs that might be relevant to Task 2, nor culled the agency-identified programs, even though some extend beyond the traditional bounds of earth science and some are loosely connected to research or training. Managers for each program provided information requested by the committee, including the size and scope of the program, goals, successes, and methods for evaluating success and for building participation of underrepresented groups.

The federal programs and some nonfederal programs were discussed in a 2-day workshop attended by 40 experts, including managers of earth science education and outreach programs and individuals knowledgeable about education, the transition into earth science careers, and program evaluation. Workshop presentations and breakout groups focused on criteria for evaluating programs (Task 3), factors required for programs to succeed (Task 4), and successes and problems in increasing diversity (Task 6). Additional information for Task 6 was gathered from agency responses to a committee questionnaire about current and potential agency partnerships and barriers to leveraging resources. The committee used the workshop results, along with published articles and reports and the committee's own knowledge and experience, to address Tasks 2, 3, 4, and 6.

Task 5 concerns a different aspect of earth science education: the knowledge and skills needed by earth scientists in their careers. In keeping with the charge, the committee confined its discussion to results from NRC earth science workforce reports, as described below.

EARTH SCIENCE KNOWLEDGE AND SKILLS IDENTIFIED IN NRC WORKFORCE REPORTS

Only two NRC workforce reports contain information on knowledge and skills needed for earth science careers: *Emerging Workforce Trends in the U.S. Energy and Mining Industries: A Call to Action* (NRC, 2013a), which covers oil and gas, mining, and geothermal energy occupations; and *Future U.S. Workforce for Geospatial Intelligence* (NRC, 2013b), which covers geospatial occupations. Both reports focus on the current and future availability of experts in subdisciplines of earth science for the workforce and only touch on the knowledge and skills required for jobs in these subdisciplines.

Future U.S. Workforce for Geospatial Intelligence (NRC, 2013b) examines 10 subject areas that underpin geospatial intelligence, including geodesy and geophysics. The knowledge and skills that are important for a career in geospatial intelligence are generally taught in 4-year colleges and universities and include the following:

- Geodesy: use of mathematical tools such as least-squares adjustment, Kalman filtering, and spectral analysis; the principles of gravity field theory and orbital mechanics; the propagation of electromagnetic waves; and the theory and operation of observing instruments such as Global Navigation Satellite System receivers and inertial navigation systems

- Geophysics: mathematical training; the principles of physics; geodesy, seismology and the structure and evolution of the Earth, including plate tectonics; the theory and measurement of the Earth's magnetic field; and space physics

The report also summarizes tiers of skills identified by the Department of Labor for the geospatial technology industry. These skills are primarily basic (e.g., interpersonal skills, effective communication, creative thinking), but also include positioning skills needed for jobs in geodesy.

Emerging Workforce Trends in the U.S. Energy and Mining Industries: A Call to Action (NRC, 2013a) examines seven industries, including the oil and gas, mining, and geothermal energy industries. Jobs in these industries generally require some college, but not necessarily a bachelor's degree. The report does not discuss specific knowledge or skills needed for jobs in these industries, but notes that many energy and mining jobs require a strong foundation in STEM—including applied mathematics, reading for information, and locating information—and that people with these skills are hard to find. It also describes efforts to define STEM skills for the energy industry through, for example, competency models and certificate programs.

ORGANIZATION OF THE REPORT

This report examines 25 federal earth science education and training programs, lays out a conceptual framework for thinking about how these programs fit together, and suggests ways to leverage federal resources to improve recruitment of a diverse population into earth science pathways. Chapter 2 summarizes the legislative authorities of federal agencies for STEM education and provides a brief overview of the federal earth science education and training programs considered in this report. Chapter 3 describes how these diverse programs can be linked to move students through informal and formal education toward an earth science career. It also discusses the critical incidents that can lead students to enter or exit the field. Chapter 4 summarizes principles for evaluating programs and shows how to use these principles and the conceptual framework to evaluate the success of earth science education and training programs. Finally, Chapter 5 describes steps federal agencies can take to increase the participation of underrepresented groups in earth science, particularly women and minorities. Supporting information for the chapters is provided in appendixes. Appendix A cites the legislative authorities for federal STEM education programs. Appendixes B and C contain the agenda and participants list, respectively, for the September 2012 workshop on earth science education and training programs. Appendix D summarizes evaluation information provided by managers of the 25 education programs. Biographical sketches of committee members appear in Appendix E, and a list of acronyms and abbreviations is given in Appendix F.

2

Federal Earth Science Education and Training Programs

Most federal agencies are involved in science, technology, engineering, and mathematics (STEM) education, commonly to satisfy legislative mandates, to support their mission, or to build a pool of potential recruits. This chapter summarizes the legislative authorities for STEM activities held by federal agencies with substantial earth science programs (Task 1) and describes federal earth science education programs that have a research or training component (Task 2). Agencies with relevant earth science programs include the U.S. Geological Survey (USGS), National Science Foundation (NSF), Department of Energy (DOE), National Aeronautics and Space Administration (NASA), U.S. Department of Agriculture (USDA), Environmental Protection Agency (EPA), National Oceanic and Atmospheric Administration (NOAA), National Park Service (NPS), and the Smithsonian Institution.

LEGISLATIVE AUTHORITIES FOR STEM EDUCATION

Legislative authority to support STEM initiatives varies widely among federal agencies (see Appendix A). Agencies with clear authority include NASA and NOAA, which were required by the America COMPETES Reauthorization Act of 2010 (Public Law 111-358) to "carry out and support research based programs and activities designed to increase student interest and participation in STEM." NSF's legislative authority, which dates back to the agency's establishment, is "to initiate and support basic scientific research and programs to strengthen scientific research potential and science education programs at all levels in the mathematical, physical . . . and other sciences" (National Science Foundation Act of 1950, Public Law 81-507). EPA is authorized to "develop and support programs to improve understanding of the natural and built environment and the relationships between humans and their environment" (National Environmental Education Act of 1990, Public Law 101-619). The Smithsonian Institution's mandate is perhaps the most expansive—it was created "for the increase and diffusion of knowledge among men" (Act of August 10, 1846, 9 Stat. 102).

Compared to these broad mandates, the Department of the Interior (DOI) has much less legislative authority for STEM education and it is related to youth conservation programs. The Youth Conservation Corps Act of 1970 (Public Law 91-378) established a program for young adults between ages 15 and 18 to perform tasks on lands and waters administered by USDA and DOI. The Public Lands Corps Act of 1993 (Public Law 91-378, as amended by Public Law 103-82) established a federal corps of young adults to work on conservation projects on federal, Indian, and Hawaiian homelands in exchange for living expenses and educational benefits.

Only a few federal agencies have specific authority for earth science education. The National Geologic Mapping Act of 1992 (Public Law 102-285) and subsequent reauthorizations provide for USGS-sponsored education in geologic mapping and field analysis. DOE has been directed to promote education and training in methane hydrate resources (Methane Hydrate Research and Development Act of 2000, Public Law 106-193) and to support education and outreach activities in energy science-related fields (Energy Policy Act of 2005, Public Law 109-58). NASA is authorized to fund museum and planetarium programs related to fields in its purview, including earth science (NASA Authorization Act of 2005, Public Law 109-155).

FEDERAL EDUCATION PROGRAMS CONSIDERED IN THIS REPORT

Each of the federal agencies with significant earth science programs was asked to identify and describe its earth science education programs that have a research or training component, either formal or informal. The programs identified by the agencies are summarized below and discussed in subsequent chapters. The descriptions reflect the status of the programs in 2012.

U.S. Geological Survey

National Cooperative Geologic Mapping Program (EdMap)—The program seeks to educate students in proper geologic mapping and interpretation techniques, basic earth science principles, and the scientific method. It provides funding to geology professors to engage upper-level undergraduate and graduate students in geologic mapping projects. The program funds approximately 62 students per year and had a budget of about $470,000 in fiscal year (FY) 2012. More than 1,000 students have gone through the program since its inception 17 years ago.

Cooperative Summer Field Training Program (in collaboration with the National Association of Geoscience Teachers [NAGT])—In this program, NAGT solicits nominations of outstanding students from field camp directors and the USGS matches candidates with available scientists in USGS research units. The goal is to partner a highly able intern with a quality science mentor to work on a meaningful earth science research project. Created in 1965, the program funds 50 students per year and had a budget of $400,000 in FY 2012.

Youth Internship Program—The program provides work experience through two programs: the Student Temporary Employment Program and the Student Career Experience Program. In the latter program, which is available to students at the high school to graduate level, the work experience is directly related to the students' academic field of study. The program funds 185 students per year and had a budget of $900,000 in FY 2012. The program has been in place for 3 years.

Hydrologic Technician Internship Program—The program seeks to stimulate ongoing interest in water science among college undergraduates and to build a pool of well-prepared new college graduates to fill vacancies at the USGS. Students are paired with an agency scientist for 10 weeks at a USGS facility. The program funds 15 students per year and had a budget of $75,000 in FY 2012. The program has been in place for 3 years.

National Science Foundation

Earth Sciences Research Experience for Undergraduates (REU) Program—The program supports research by undergraduate students in any scientific area funded by NSF's Division of Earth Science. The research may be part of an ongoing program or a project designed specifically for the REU Program. The program funds approximately 215 students per year and had a budget of $1,500,000 in FY 2012. The REU Program has existed for more than 20 years, and each REU site is funded for 3–5 years.

Opportunities for Enhancing Diversity in the Geosciences (OEDG) Program—The goals are to increase participation of African Americans, Hispanic Americans, Native Americans, Native Pacific Islanders, and persons with disabilities in earth science, and to increase the perceived relevance of earth science in underrepresented groups. Typically, NSF receives 80–100 proposals and funds about 35 percent of them. The program, which was created in 2002, funds about 14,000 students per year and had a budget of $3,600,000 in FY 2012.

Earth Sciences Postdoctoral Fellowships—The goal of the program is to provide early-career investigators with research and education experience that will help them establish leadership positions in the scientific community. Applicants submit proposals to carry out a research project and an education activity for 2 years at an institution of their choosing. Created in 2008, the program has funded at least 10 fellows per year at $85,000/year for each fellowship.

Geoscience Education (GeoEd) Program—The program supports projects to improve formal and informal earth science education, to increase the number of students pursuing earth science, to broaden participation of underrepresented groups, and to engage the public in Earth system science. The number of students varies by project, and the program's budget was $1,500,000 in FY 2012. The program began in 1997.

Global Learning and Observations to Benefit the Environment (GLOBE) Program; in collaboration with NOAA and NASA)—The program connects students, teachers, and scientists through inquiry-based investigations of the Earth system. Program goals include improving student understanding of environmental and Earth system science, building a global community, and engaging the next generation of scientists and global citizens in activities to benefit the environment. The program has involved about 1.5 million students in approximately 24,000 schools since its inception in 1994. NSF's contribution to the budget was $1,100,000 in FY 2012.

Geoscience Teacher Training (GEO Teach) Program—This one-time competition in 2006 funded two programs designed to improve the quality of middle school and high school instruction in earth science. GEO-Teach focused on providing teachers with curricular materials and preservice teacher training, and creating in-service professional development programs to enhance students' understanding of earth science. The program funded approximately 2,000 preservice teachers and had an annual budget of $2,000,000.

Department of Energy

Relevant programs are offered at the agency, national laboratory, facility, and research project levels. Examples are described below.

Office of Science Graduate Fellowship (SCGF) Program—The program provides 3 years of support to students pursuing graduate training in basic research in fields of study relevant to DOE's Office of Science, including earth science. The ultimate objective is to encourage the development of the next generation of scientific and technical talent in the United States. The program was established in 2009 and supported 150 fellows in 2010. The budget for FY 2012 was $5,000,000.

Summer of Applied Geophysical Experience (SAGE) Program—The program introduces students to field methods in geophysical exploration and basic and applied research through a 6-week course. The program funds 20–25 undergraduates and 4 or 5 graduate students per year. SAGE was established in 1983 and had a budget of $120,000 in FY 2012.

Science Undergraduate Laboratory Internships (SULI) Program—The program encourages undergraduate students to pursue STEM careers by providing research experiences at DOE laboratories, as well as providing professional development workshops and scientific lectures and seminars. Internships last 10 or 16 weeks and focus on projects that support the DOE mission. The program, which was established in 1999, has grown to support 700 interns per year and had a budget of $6,500,000 in FY 2012.

Community College Internships (CCI) Program—The program provides community college students interested in a technical career with technical training experiences and professional development activities at DOE laboratories. Students spend 10 weeks working on technologies, instrumentation projects, or major research facilities that support DOE's mission. CCI supports nearly 100 interns per year and had a budget of $600,000 in FY 2012. It was established in 2001.

National Aeronautics and Space Administration

Gravity Recovery and Climate Experiment (GRACE) programs—This education and public outreach program is associated with a satellite that measures the Earth's gravity field. The goals include increasing teacher and student understanding of Earth's history, Earth system science, and global climate change. The education component began in 2011, although the GRACE satellite has operated since 2002. The program funded 1,585 students and had a budget of $80,000 in FY 2012.

U.S. Department of Agriculture

Agriculture and Food Research Initiative (AFRI) projects—The program funds research, education, and extension grants that address issues important to sustaining agriculture, including renewable energy, natural resources, and environment. The education and extension efforts are intended to provide scientific knowledge needed for people to make informed practical decisions. The program, which began 2 years ago, currently funds about 500 undergraduate and graduate students and postdoctoral fellows and had a budget of $25,000,000 in FY 2012.

AFRI National Institute of Food and Agriculture Fellowships Grant Program—The program trains students in agricultural, forestry, and food science. The goals include strengthening the ability of the scientific community to meet challenges facing agriculture, forestry, and food systems; developing technical and academic competence of doctoral candidates; and strengthening research and teaching of postdoctoral scientists. The program funds 54 fellowships per year and had a budget of $12,000,000 in FY 2012. The first fellowships were awarded in 2011.

4-H Environmental Education/Earth Science programs—The program aims to increase science awareness, skills, and knowledge among youth and to increase awareness of opportunities to contribute to society using science skills. Programs are developed in partnership among the National Institute of Food and Agriculture, USDA, and state land grant colleges. Total enrollments for the programs were 1,390,553 in FY 2010, and federal support for the entire 4-H Program was approximately $47,000,000.[1] These programs date to the inception of 4-H in 1902.

[1] Estimate from the USDA Cooperative Extension System.

Environmental Protection Agency

Greater Research Opportunities (GRO) Undergraduate Fellowship Program—The program supports undergraduate students in environmental science fields for their last 2 years of study and provides an EPA internship. The program goal is to increase the number of environmental scientists, engineers, and policy experts in the U.S. workforce. The program has historically funded 40 students per year and had a budget of $2,000,000 in FY 2012. The program has existed since the early 1980s.

Science to Achieve Results (STAR) Graduate Fellowship Program—The program supports master's and doctoral candidates in traditional and emerging disciplines of environmental science in order to increase the number of environmental scientists, engineers, and policy experts in the U.S. workforce. The program, which began in 1995, has historically funded 80–100 students per year and has had an annual budget between $3,400,000 and $4,500,000.

National Oceanic and Atmospheric Administration

Educational Partnership Program (EPP) with Minority Serving Institutions (MSIs)—The goal is to increase the number of trained graduates, particularly from underrepresented communities, in STEM fields directly related to NOAA's mission; and to strengthen collaborative research between NOAA scientists and researchers at MSIs. EPP components include scholarships and internships at NOAA facilities for undergraduate and graduate students, as well as competitive awards for Cooperative Science Centers to build capacity in mission areas at MSIs. The program funds an average of 300 students per year and has had an average annual budget of $14,000,000. The program has been in existence for 12 years.

National Park Service

Geoscientists-in-the-Parks Program (in partnership with the Geological Society of America)—The program places interns in parks to carry out research and monitoring and to provide earth science interpretation and education and outreach assistance to park managers and staff. More than 100 interns, mostly between the ages of 18 and 26, gain earth science work experience in the program each year. The annual budget varies and was approximately $740,000 in FY 2012. The program began in 1996.

Geoscience-Teachers-in-Parks Program (in partnership with NAGT)—Created in 1996, the program seeks to exchange learning and scientific research between the park, local earth science teachers, and communities; to advance educational and interpretive opportunities at the park; and to develop lifelong networks with local communities, schools, and the park. Two or three teachers participate every year. The annual budget is variable and was $12,000 in FY 2012.

National Fossil Day—This annual event is aimed at promoting public awareness and stewardship of fossils and increasing appreciation of their scientific and educational value. Hundreds of individual events and activities are hosted by more than 240 partners throughout the United States. The program reached about 15 million students in 2011 and had a budget of $40,000 in FY 2012. The first National Fossil Day was held in October 2010.

Smithsonian Institution

Leadership and Assistance for Science Education Reform (LASER) Program—The goal is to improve science education programs in U.S. schools through a K–8 inquiry-centered curriculum. LASER regions serve as focal points for building on previous accomplishments in regional K–8 science education reform. The program reaches school districts and states representing 30 percent of the K–8 population and has an annual budget of $3,000,000 to $6,000,000. LASER was launched in 1998.

SUMMARY

Most federal earth science agencies have authority for STEM education, although the scope varies among agencies. NASA, NOAA, and NSF have the broadest authorities for STEM initiatives; DOI authority is limited to the establishment of youth conservation programs. Only a few agencies (i.e., NASA, DOE) have specific authority for earth science education. Nevertheless, most agencies with an earth science purview (e.g., USGS, NSF, DOE, NASA, NOAA, EPA, USDA, NPS, Smithsonian Institution) have developed earth science education and training programs. The programs vary widely in size, ranging from a few to thousands of participants per year, and in goals and objectives, as discussed in the next chapter.

3

A Program Framework

The second task of the committee was to examine federal earth science education programs with a research or training component. The programs, which are summarized in Chapter 2, are commonly aimed at a specific goal, such as attracting underrepresented groups to earth science, teaching laboratory or field skills, or providing job experiences. As pointed out at the workshop, however, they can also contribute to the larger goal of building the earth science workforce (Box 3.1). This chapter describes a framework for thinking about federal earth science education and training programs in the context of a larger system that moves individuals from inter-

BOX 3.1 Workshop Discussions on an Education Program Framework and Critical Incidents

Key points raised by individuals at the workshop included the following:

- The need for a community model or framework of programs that engages students and leads them from awareness to employment in earth science.
- Understanding critical incidents that persuade individuals to enter earth science (e.g., hiking with families, nature books, earth science classes) or cause them to leave it, which can inform creation of a program framework.
- The importance of connecting programs along the pathways to earth science careers, offering multiple and varied education opportunities, and facilitating students' movement along the pathways.
- Ways to take advantage of the unique aspects of earth science (e.g., connection to land and place, adventurous or outdoors nature, systems thinking) to engage and create a science-literate public.

est to employment in earth science. The paths of individuals through the system vary, as do their entry and exit points. The chapter concludes with a discussion of critical incidents: the events and influences on people's lives that lead them to pursue particular educational paths toward a career.

FRAMEWORK CONCEPT

Preparing a student for the science, technology, engineering, and mathematics (STEM) workforce commonly takes place over years or even decades and requires training and experience from a variety of programs, institutions, and individuals. Individuals become interested in a STEM discipline in different ways and at different times. Once that interest is sparked, it must be nurtured through opportunities to explore and learn about the discipline, to develop and practice skills, to obtain guidance, and to investigate job opportunities. The particular paths of individuals from interest to employment will depend on factors such as their specific interests, the educational and workforce opportunities available to them, and the needs and expectations of their families (Lent et al., 1994, 2000, 2008; Houlton, 2010; Maltese and Tai, 2010, 2011). Thus, preparation of the future workforce can be thought of as a system of opportunities and experiences that link together in ways that enable individuals to move from their own entry point through a series of experiences that prepare them for employment that they will find rewarding, and ultimately into the workforce.

This system of opportunities and experiences in earth science is illustrated in Figure 3.1, which was developed based on the experience of committee members and workshop participants. In this framework, individuals first become aware of earth science, then engage in learning the field, and eventually prepare for a career by acquiring specialized knowledge, skills, and expertise and by exploring different employment options. The framework is portrayed as a triangle because more individuals will develop an interest in earth science than will become engaged in the field, and more will become engaged in the field than will pursue professional preparation and employment.

Different types of education and training opportunities are represented by the upward-pointing polygons in Figure 3.1. Programs with multiple goals or audiences can span more than one stage of the framework (e.g., classes raising awareness of earth science may be taught at elementary to college levels). The programs may also play multiple roles because individuals bring different goals and experiences to the opportunity (e.g., a research opportunity may be used by students to prepare for the profession or to obtain skills and recommendations needed for graduate school). Figure 3.1 shows where these earth science education programs are commonly placed; the exact placement of a particular program will depend on its target audience and goals.

Of course, the paths that individuals actually take are often more complicated than is implied in Figure 3.1. The pathway from interest to employment can be full of twists, turns, and detours, and individuals may enter or leave the path at different points. Moreover, Figure 3.1 does not show a critical third dimension: the specific area of earth science in which the student specializes. As students progress upward to prepare for employment, they also move laterally along this third dimension as their interests develop. For example, a student may first become interested in earth science through an introductory course on the geology of national parks. As she learns more about the earth sciences, she might become particularly interested in the hydrology of water resources and begin to prepare for graduate work in this area. A summer research experience in this field might confirm this interest or reveal a stronger interest in the geochemistry of water. After a master's program and internship, she may embark on a career focused on cleanup of mine drainage. Connections between opportunities and movement in all three dimensions are integral to creating the range of skills and expertise needed by the modern workforce.

Although this report focuses on high school and college programs that prepare students for the workforce, the full framework is outlined because programs for older students depend on pro-

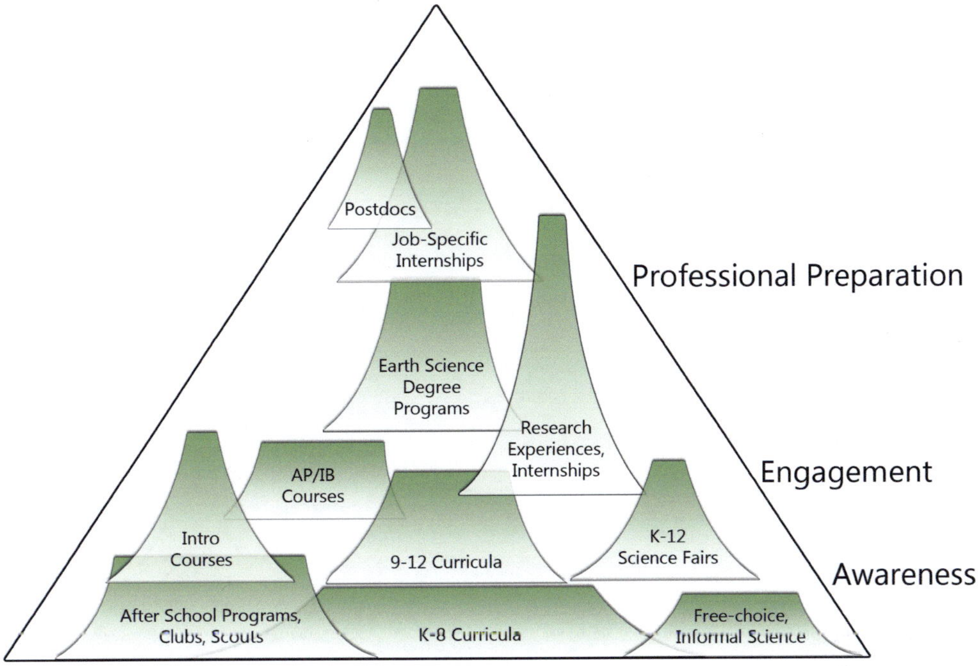

FIGURE 3.1 Conceptual framework illustrating the types of programs and experiences (tapering polygons) that help move individuals along a pathway from awareness of earth science (base of the triangle) to the earth science workforce (apex of the triangle). Relevant programs include those provided as part of a student's formal education, educational programming offered outside of the formal system, and informal learning opportunities. Polygons are not drawn to scale, but their vertical extent is intended to show that some education opportunities span more than one stage of the framework and their relative horizontal extent is intended to show that more individuals participate in awareness activities than in professional preparation activities.

gramming at lower levels. The general stages that students follow through the system are described below.

Federal Programs in the Context of the Framework

Federal education and training programs contribute to all stages of the framework, from interesting students in earth science to education and outreach programs to providing internships, traineeships, and research opportunities within federal agencies. Figure 3.2 shows the federal earth science education and training programs described in Chapter 2 in the context of the committee's framework. The roles of these programs in the various stages of the framework are described below. Although many of the programs span more than one stage of the framework, each is given as an example only once below.

Awareness

Awareness arises from activities that bring earth science to the attention of an individual. A robust set of educational pathways includes a diversity of mechanisms for bringing earth science awareness to the widest possible spectrum of individuals—including K–12 and undergraduate students, and parents and other adults—at different times. Converting awareness to interest may

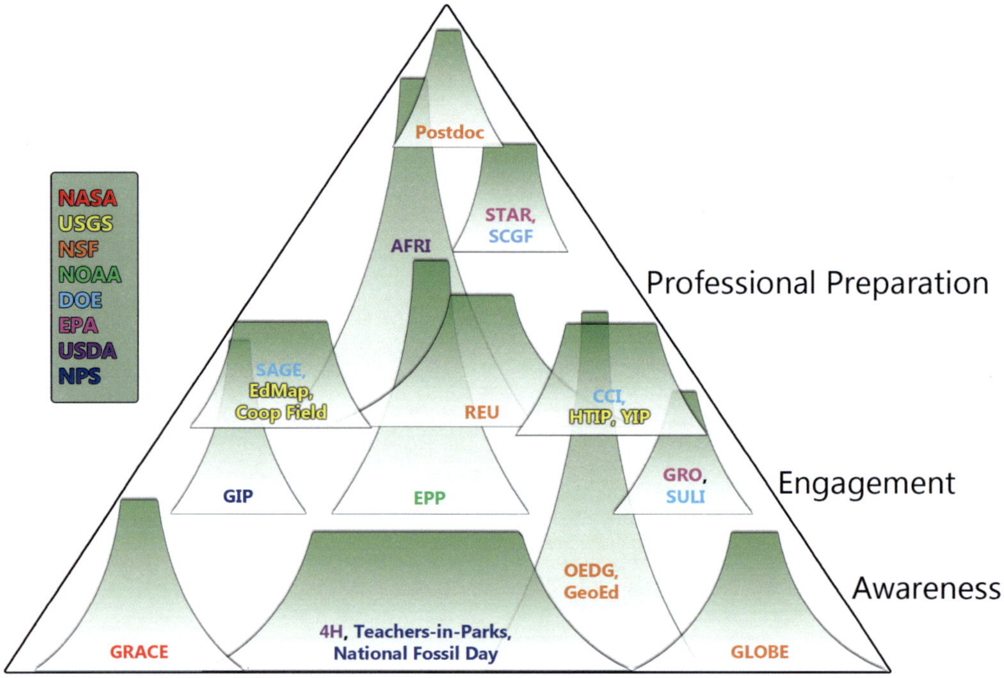

FIGURE 3.2 Placement of the federal earth science education and training programs considered in this report in the committee's conceptual framework. Because of space constraints, programs that have a small earth science component (Smithsonian LASER program) or that were one-time competitions (National Science Foundation's GEO-Teach Program) are not shown.

require a series of positive experiences. Productive mechanisms for developing awareness include introducing earth science concepts in formal education; informal learning in museums, after-school programs, and clubs; and individual exploration through books or other media. Informal learning is particularly important for building awareness of earth science, which is not widely taught in school (Underwood, 2008; Windschitl et al., 2008). Because parents are gatekeepers to young students' access to informal education, earth science programs for families are important.

Federal agencies play a major role in developing earth science awareness. The U.S. Department of Agriculture's 4-H club offers activities that span a wide range of topics, including earth science, to large populations of youths. The National Park Service organizes National Fossil Day, which focuses attention on fossils in schools, in informal settings, and through the media. Its Teachers-in-Parks Program engages teachers in developing awareness-building materials for use both in the park and in school districts nearby.

Engagement

In the engagement stage, students actively engage in learning about the Earth and earth science by choosing earth science activities or study. Engagement can be fostered through activities that are relevant to students, that give them a sense of contributing to their community or to society at large, that engage them in solving problems they find interesting, or that allow them to synthesize and make use of prior learning (PCAST, 2012). Middle school, high school, and college are important venues for fostering engagement. In these settings, a student has an opportunity to learn about the discipline in a structured way. Activities such as science fair projects, course projects, or service learning projects support students' active engagement in problem solving using earth science methods and habits of mind. Shadowing earth science professionals or doing internships provides opportunities to build understanding of the profession as well as to develop an identity as a future earth scientist.

Federal agencies offer a variety of engagement programs and also play an indirect role in engagement by providing public access to their data. The U.S. Geological Survey (USGS) Youth Internship Program helps students learn more about earth science through internships and project-based learning. The National Science Foundation's (NSF's) GEO-Teach Program funded the development of curricula aimed at increasing the use of data-rich activities and pedagogies that engage students in formal education at all levels. Projects supported by NSF's Opportunities for Enhancing Diversity in the Geosciences program include those aimed at expanding the participation of underrepresented groups in earth science. The GeoFORCE program, funded partly by the USGS, engages high school students in mentored, immersive summer field experiences.

Professional Preparation

The transition from engagement to professional preparation is not a distinct event but rather a shift in an individual's education strategy and focus. It occurs when an individual changes his or her view from an exploration of earth science to the acquisition of knowledge, skills, abilities, and professional attitudes needed for a particular type of job. The path may be short (e.g., an adult returning to school for new professional training) or long (e.g., a student exploring various interests, settling into a course of study, and then seeking a job). At this stage, students with different employment goals will select different programs or different electives within a major. Club activities, cohort groups, speakers, professional society meetings, career counseling, and mentoring help students identify and then obtain the expertise, confidence, and other professional attributes they will need in their desired job. Research experiences and internships allow students to explore their interest in a particular aspect of the field, to build data collection and analysis skills, and to develop higher-order thinking skills and expertise in a specialty area. Undergraduate and graduate internships and postdoctoral positions introduce students to job opportunities and employers and help crystallize work abilities, interests, and values. Programs that attend to both cognitive and affective skills are particularly important for underrepresented groups (e.g., Jolly et al., 2004; NRC, 2011).

Federal agencies play a key role in offering research opportunities for undergraduate students. NSF's Research Experience for Undergraduates program funds the participation of undergraduates in research projects. The Environmental Protection Agency's Greater Research Opportunities Undergraduate Fellowship Program and the Department of Energy's (DOE's) Science Undergraduates Laboratory Internships bring students to work with scientists at agency facilities. The USGS Cooperative Field Training Program and DOE's Summer of Applied Geophysical Experience Program allow undergraduates to work on hands-on projects with agency scientists.

Federal agencies also offer internships and transitional employment opportunities for undergraduate, graduate, and postdoctoral students. For example, the USGS Hydrologic Technician Internship Program and DOE's Community College Internships Program are aimed at drawing future technicians from the 2-year college system, and the National Cooperative Geologic Mapping Program is aimed at developing field mapping expertise, a critical skill for the USGS scientific workforce. These programs target different educational levels, areas of the country, and skills in ways that will help move students into different parts of the federal workforce.

DEVELOPING A SYSTEM APPROACH

The federal earth science education and training programs described above operate largely in isolation from one another, making it more difficult for program managers to find gaps among the collective programs. Moreover, the burden of finding a path from one opportunity to another rests heavily on students, who have only limited knowledge and experiences to draw on, and their advisors, who are most aware of opportunities and pathways in their own specialties. As a result, opportunities are missed to entrain and retain talented students.

Increasing the size and diversity of the earth science workforce requires not only a variety of education and training programs, but also interactions among programs to support the movement of students from interest to employment. Connecting educational and training opportunities to employers would enable the system to be responsive to changing workforce needs. For example, growth of the solar power and electronics industries, which depend on particular rare earth elements and metals, could increase demand for economic geology graduates.

Networks are a primary mechanism for connecting diverse activities, programs, and organizations. Some types of earth science networks are already in place. For example, major petroleum companies tend to recruit at a particular set of academic institutions, particularly in the Gulf Coast area. At the national level, professional societies such as the Geological Society of America and the American Geophysical Union connect students with potential employers and graduate schools across the country by advertising job openings and providing a venue for interviews. The American Geosciences Institute provides online resources for students and families, describing career pathways for earth scientists. Although these networks are useful, stronger and more systematic connections are needed to create synergies between programs and to support the movement of students through the system of opportunities.

Opportunities that interest students in local jobs are valuable. However, the nature of the earth science workforce varies by region. Lack of national networks can limit student opportunities by making it difficult for them to move from education in one part of the country to employment in another, or to find jobs in specialties that are not represented in their local educational institution or community. For example, if all petroleum geologists are trained in Texas, earth science students in New England are unlikely to know much about the oil and gas industry. This geographic focus limits not only the potential workforce, but also the ability of students to appreciate the needs, challenges, and contributions of the various earth science specialties.

A connected system of opportunities is particularly important for attracting and retaining students from underrepresented groups (NRC, 2011). Jolly et al. (2004) found that a combination of engagement (defined as awareness, interest, and motivation to study the field), capacity (knowledge and skills), and continuity (institutional and programmatic opportunities, material resources, and guidance) is necessary to keep underrepresented students on a path to a science career. They also found that access to networks is the key to continuity. Cultural and ethnic affinity organizations (e.g., SACNAS, American Indian Science and Engineering Society, National Association of Black Geoscientists) and other groups are addressing this challenge by building networks for underrepresented students. However, a system of earth science education and training opportunities would benefit all students.

Linkages between the various education opportunities can be strengthened by increasing the visibility and person-to-person connections between programs. For example, a more systemic approach to advertising education and training opportunities (e.g., by creating a central listing of available internships) and illustrating educational pathways to employment (e.g., by expanding competency frameworks)[1] could improve the ability of students to navigate through programs. Connecting federal programs to 2-year colleges, which play a key role in preparing students for a

[1] See http://www.careeronestop.org/CompetencyModel/pyramid.aspx.

bachelor's degree, would also help students move along earth science pathways. Increasing opportunities for program leaders to interact with one another could strengthen connections between programs. Establishing such mechanisms would require the collective effort of federal agencies, professional societies, nongovernmental organizations, educational institutions, and employers.

CRITICAL INCIDENTS AND PATHWAYS THROUGH THE FRAMEWORK

Students enter and leave earth science at different points and for different reasons. The specific events that lead people into certain career and educational paths are commonly referred to as critical incidents. The idea was pioneered by Flanagan (1954) and has been formalized into a social science research methodology known as the critical incident technique.

A handful of studies have investigated the specific factors and pathways by which students discover and pursue formal education in earth science. Levine et al. (2007) and Houlton (2010) identified a series of specific incidents, decision moments, or events that strongly influenced students to choose earth science as an undergraduate major and career path. These studies involved relatively small samples: 17 earth science students in the midst of their undergraduate education at two major midwestern research universities (Houlton, 2010) and 14 earth science faculty members and other professionals who are also ethnic minorities (Levine et al., 2007). The subjects of the latter study were from diverse backgrounds and were educated in a wide array of universities in the United States. The goal of these studies was to carry out a detailed and relatively deep qualitative analysis of the common pathways and significant moments that lead people into earth science, not to provide a statistically significant, generalizable model of earth science career choices and critical incidents. Nevertheless, the results of the two studies were consistent, potentially revealing some broad outlines of common career pathways in earth science.

Figure 3.3 illustrates the general pathway from interest to education to employment in earth science, which can be seen as parallel to a student's journey through the conceptual framework

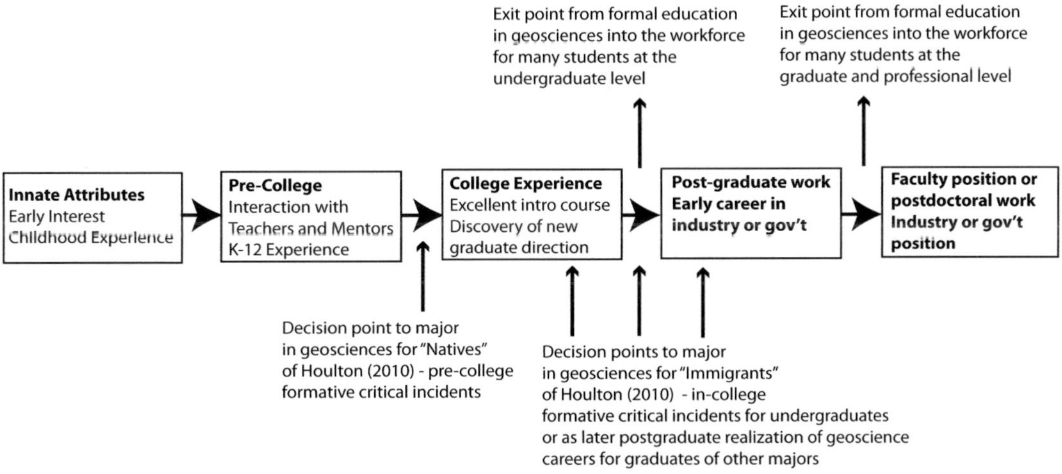

FIGURE 3.3 Schematic illustration of the path from informal and formal education in earth science to a career in academia, industry, or government (boxes connected by horizontal arrows). Critical incidents (listed in the boxes) draw some students to the path relatively late (text at the bottom of the figure). Exits to an earth science career can be made from multiple places along the path (right two boxes and text at the top of the figure), depending on the degree requirements of the position. SOURCE: Modified from Houlton (2010).

discussed above. Students may enter the path at several points, often prompted by one or more critical incidents that raise awareness of earth science or increase interest and engagement in the field. Similarly, there are multiple exit points from formal education to an earth science career, depending on the degree requirements of the position, which range from associate's degrees for technicians to doctorates for researchers.

Entry to Earth Science

Results from Levine et al. (2007) and Houlton (2010) suggest that the pathway into earth science differs for two populations: (1) individuals who reach the awareness and engagement stages before they reach college ("natives" in Figure 3.3); and (2) individuals who find their way to the field much later, in college or even in postgraduate work ("immigrants" in Figure 3.3). Many of these latecomers discover the field through an outstanding introductory earth science course taken during their undergraduate work. This result echoes the widespread anecdotal experience of earth science faculty, who often characterize geology, geophysics, and related fields as discovery majors. The strongest influences on students' choice in the major are friends, family, schoolteachers, and previous courses; weaker influences include faculty advisors, other faculty members, and professional advisors (Hoisch and Bowie, 2010).

The Houlton (2010) study found that certain critical incidents raise awareness of earth science or prompt decisions to enter the field, and that these incidents differ for the two populations. For students in the study who became interested in earth science early in life, an innate interest was fostered by incidents such as family trips to geologically interesting locations or by personal experience of a natural disaster. Nearly all of the critical incidents for this group were extracurricular in nature, perhaps reflecting the relatively sparse treatment of earth science in the K–8 science curriculum (Windschitl et al., 2008; AGI, 2012) and the absence of earth science courses in many high schools.

In college, students with prior interest in earth science and those who later discovered the field were drawn to pursue earth science degrees by outstanding introductory courses with integrated lab or field experience (akin to moving from awareness to engagement in the framework). A common ingredient for all students in the Houlton (2010) and Levine et al. (2007) studies was that these courses were taught by talented, energetic, and engaging instructors. These outstanding introductory courses may be a major gateway for the field, potentially at the undergraduate and graduate level. Once in an earth science major, faculty engagement in the classroom and the field continued to be important for retaining students in the discipline and encouraging them to pursue a career. Frequent access to information about pathways toward the workforce, in the form of course material, internships, or research opportunities, was also important. In the context of the framework, these types of interactions move students from engagement to professional preparation.

Exits from Earth Science

Critical incidents and other factors may discourage students from continuing study and a career in earth science. Although studies of reasons for leaving earth science are sparse, analyses of other scientific and engineering disciplines have identified a wide variety of factors that contribute to students leaving the field, such as inadequate precollege preparation, lack of funding for research involving students, negative classroom experiences with peers or faculty, inadequate faculty advice or support, and a culture of competition (e.g., Seymour and Hewitt, 1997).

SUMMARY AND CONCLUSIONS

Federal earth science education and training programs can be considered in the context of a framework that introduces individuals to earth science and then engages them in learning the discipline and acquiring the specialized knowledge, skills, and expertise they will need for an earth science career. Although the actual path from awareness to employment is often more complex, this framework is useful for organizing the various types of learning opportunities and their intended outcomes.

A rich variety of federal earth science education and training programs exists at every stage of the framework. The federal programs are usually developed and run in isolation, but connecting them into a system would serve agency needs as well as foster development of the workforce. Connecting federal programs and complementary programs offered by other organizations with employers would help the system respond to changing workforce needs. Stronger and more visible connections between programs would help students find a path to an earth science career. Such networks are particularly important for attracting and retaining underrepresented groups in earth science.

Critical incident analysis may offer insight on the types of programs that attract and retain students in earth science. A few studies on critical incidents in earth science studies suggest that two populations enter earth science pathways: (1) those who discover an interest in earth science before they reach college, commonly through extracurricular activities; and (2) those who become interested in college, often through an outstanding introductory class. These two populations of potential earth scientists pose recruitment and retention challenges because awareness and engagement opportunities must be pitched to both precollege and college students. Students exit from formal education in earth science for different reasons and at different times. The multiple entry and exit points along earth science pathways underscores the need for a wide range of education and training programs that can attract different populations of potential earth scientists at different times, and for linkages between these programs that facilitate the movement of students into earth science careers.

4

Program Evaluation

Resources for federal earth science education and training programs are generally limited, so it is important for agencies to invest in programs that work. Program evaluation provides a means for determining whether a program is succeeding and why. However, only a few of the education and training programs considered in this report have been formally evaluated or are structured in a way that facilitates evaluation, making it difficult to address Task 3 (identify successful programs) or Task 4 (determine what made these programs successful) as formulated. This chapter describes effective methods for evaluating programs, the limitations of evaluation approaches used in the federal earth science education and training programs considered in this report, and evaluation of these programs in the context of the Chapter 3 framework of education and training opportunities. Evaluation at each stage of the framework is illustrated with examples of effective practices, drawn from the literature, workshop discussions (Box 4.1), and other sources.

USING LOGIC MODELS FOR EVALUATION

Program evaluations generally focus on understanding program goals, establishing criteria for success, and gathering data to compare program performance to the criteria for success (NRC, 2009). Both formative evaluation (done while the program is under way with the goal of improvement, usually for internal audiences) and summative evaluation (done at the end of a program to determine its worth, often to external audiences; see Scriven, 1991) are needed to help providers develop effective programs and to determine the extent to which those programs met stated goals. Logic models are commonly used in program evaluation to understand how the program is supposed to work (e.g., McLaughlin and Jordan, 1999). They define who the program is trying to reach and what it is trying to achieve, and describe how to translate program resources into near-term results and long-term impacts. Logic models are often represented graphically as shown, for example, in Figure 4.1.

The logic model consolidates information on the inputs, activities, outputs, and outcomes of the program. Inputs are the resources used, such as people, time, or exhibit space. Activities are what the program does, such as attract visitors, air on television, provide summer experiences, or teach

BOX 4.1 Workshop Discussions on Criteria for Evaluating Program Success

Key points raised by individuals at the workshop included the following:

- Success can be defined in many ways (e.g., short term vs. long term, individuals vs. cohorts vs. the organization vs. the profession).
- Criteria for success depend on the goals of the program.
- Measuring the impact of informal programs as well as the connectivity among programs and between programs and a career path is difficult.
- Suitable performance measures include both quantitative data (e.g., number of participants) and qualitative data (e.g., depth of experience) and trends.

Example criteria included the following:

- Nature of the opportunity (e.g., career relevant, culturally relevant, hands-on, real world)
- Number of participants
- Diversity of participants, partners, or the resulting workforce
- Appropriate time and effort to achieve stated goals
- Use of best practices
- Increase in participants' earth science knowledge, skills, or identity with the field
- Intervention of program at critical junctures
- Connectivity of opportunities to keep participants moving along earth science pathways
- Ability to obtain other support or partners (professional societies, private companies, universities)
- Preparation of participants for employment
- Sustainability or longevity of the program
- Ability to scale from local to regional or national interests and issues

skills. Outputs are the immediate, tangible results of the program, such as the number of visitors who viewed the exhibit or the new skills learned by students. Outcomes are the longer term changes that the program aims to achieve. Earth science education programs generally aspire to three types of outcomes: awareness, engagement, or professional preparation.

To determine whether a program has achieved its objectives, each outcome variable must be measured either for a group of individuals before and after they participate in the program or for participants and an appropriate group of nonparticipants. Many measures of baseline awareness, engagement, and professional preparation can be made, but some form of survey or pretesting is likely to be needed to assess an individual's change. For example, one might test geological knowledge before a student took an upper-level earth science course, and then measure the student's geological knowledge after that course was completed.

To determine why a program worked or did not work, rather than just whether it did, the evaluation covers the activities themselves. For example, did visitors who spent more time at a geological exhibit show greater awareness on leaving it than those who spent less time? Did it make a difference whether they participated in hands-on elements in the exhibit? Examining the organizational context of a program may also provide important insights on why some programs work and others do not. For example, are programs that work with educational standards movements in schools more effective than those that blaze their own pathways? Best practices can be developed from program activities that have been demonstrated to produce the desired outcomes.

Measuring short-term outcomes is easier than measuring long-term outcomes, but the latter are more important for determining whether a program is meeting its goals. Follow-up after an

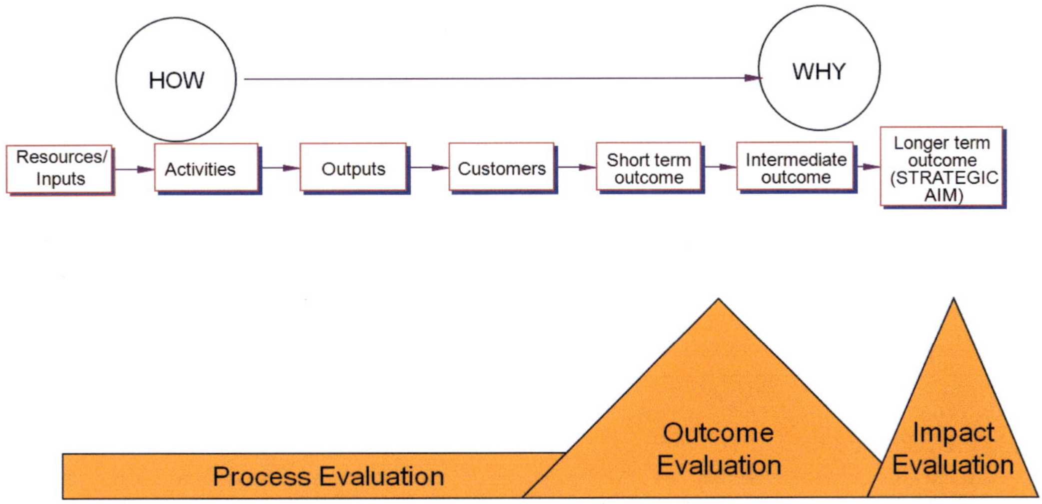

FIGURE 4.1 Example of a logic model illustrating the causal relationships among program elements (boxes) and evaluation stages (orange shapes), which show how the program works and whether and why it succeeds in generating results. SOURCE: Adapted from a 2005 presentation by Federal Evaluators (Evaluation dialogue between OMB and federal evaluation leaders: Digging a bit deeper into evaluation science), www.fedeval.net.

appropriate period of time is therefore important. Tracking individual participants over time is ideal, but even the best surveys lose track of some participants, and participants often lose interest in responding to requests for information. Surveys across similar programs may partially compensate for these problems at the level of individual programs, and they are also more cost-effective.

AGENCY PROGRAM EVALUATION

Two of the committee's tasks concern the evaluation of federal earth science education and training programs. Task 3 was to identify criteria for evaluating success and, using those criteria and the results of previous federal program evaluations, to identify examples of successful programs in federal agencies. Task 4 was to determine what made those programs successful. Important sources of information for these tasks were the workshop discussions (Box 4.1) and the written responses of program managers to the following questions:

1. What are the key goals or outcomes for the program?
2. How is the program evaluated?
3. What are the major successes of the program and what criteria are used to measure success?
4. What things have been essential to the program's success?

The answers to these questions revealed a wide range of criteria for success and evaluation approaches (see Appendix D). As noted above, criteria for success depend on the specific goals of the program. Thus, no single set of criteria can be developed to determine the success of all federal earth science education programs considered in this report. Rather, a comprehensive evaluation approach is needed to demonstrate program success.

Evaluation approaches used by the agencies range from informal assessments by an agency manager or principal investigator to rigorous external review. Few programs considered in this report have been designed to facilitate evaluation (Box 4.2) or collect the data necessary to determine whether the program succeeded or how to improve, sustain, or expand it. Even when data are collected, they are commonly not ideal for evaluation purposes. In addition, the formulation of goals and criteria for success poses problems for evaluation. Some of the stated goals are too broad to measure (e.g., improve understanding, build a community). In some cases, program goals are narrow (e.g., increase the number of participants), but the evaluation criteria are simple enumeration measures, which provide only limited information on program success. Only a few programs try to measure the impact of their program toward long-term, strategic aims (e.g., recruiting and retaining minorities, attitudes toward context-specific activities). Finally, the criteria do not always match the stated goals. For example, measuring participant satisfaction with the program does not indicate whether more students chose an earth science career. The mismatch of goals and measures confounds the ability to define program-level criteria for success.

External evaluations have demonstrated the success of the Opportunities for Enhancing Diversity in the Geosciences (OEDG) Program, the effectiveness of the selection process for Science to Achieve Results (STAR) fellows, and the progress toward achieving Educational Partnership Program (EPP) goals (Box 4.2). Other federal programs considered in this report cite successes (e.g., participants obtain earth science positions; see Appendix D), but the program information provided by the agencies was insufficient for the committee to make an independent determination. The lack of suitable evaluation data across programs underscores the importance of incorporating evaluation into the program design. By using a logic model in the context of the Chapter 3 framework of education and training opportunities, it would be possible to evaluate success at several levels: (a) whether a program is achieving its particular goals; (b) a program's contribution to increasing awareness, engagement, or professional preparation; and (c) a program's contribution to preparing a skilled and diverse workforce, including which programs work for which target groups and under which circumstances. Program evaluation in the context of the framework is described in the next section.

Because the committee lacked the robust data needed to choose successful examples of federal earth science education and training programs, it could not offer insight on why these programs are successful (Task 4). In assessing their own programs, managers identified several factors for success. The most common were stable funding, cost sharing, the commitment of agency managers or principal investigators, and partnerships. Agency support and community outreach were also important for many programs. Some managers highlighted program design—such as a good fit between participants and providers, flexibility, or institutionalization—as important for success. All of these factors are reasonable and consistent with workshop discussions, which also identified the involvement of families and the motivation for mentors (e.g., recognition of service) as important factors.

PROGRAM EVALUATION IN THE CONTEXT OF THE FRAMEWORK

Key elements of logic models (inputs, activities, outputs, and outcomes) and effective evaluation practices for programs at different stages of the framework are described below. The discussion is illustrated using exemplars that embody at least some elements of logic models in their evaluation.

Awareness

Awareness activities (e.g., formal education, after-school programs or clubs, earth science exhibits in museums) are designed to attract individuals to earth science, often through their own choice to participate. Participants include students and families of students in elementary school through high school, with researchers and scientists providing the content (inputs to the logic

BOX 4.2 Formal Evaluations of Federal Earth Science Education and Training Programs

Most of the federal earth science education and training programs considered in this report use relatively informal evaluation methods (see Appendix D). A few have undergone a more rigorous external review, either as individual programs (e.g., National Science Foundation [NSF] OEDG Program, Environmental Protection Agency [EPA] STAR Graduate Fellowship Program) or as part of a broader education portfolio (e.g., National Oceanic and Atmospheric Administration [NOAA] EPP, NSF Research Experiences for Undergraduates [REU] Program). The methods and results of these formal evaluations are summarized below.

NOAA Educational Partnership Program. A National Research Council review (NRC, 2010) found that a variety of evaluation methods were used for NOAA educational programs, ranging from no formal evaluation (e.g., EPP) to an outcome-based summative evaluation. The NRC evaluated the EPP using information provided by NOAA or gathered in interviews of NOAA staff and site visits. The evaluation found that the EPP had made progress toward achieving its goals. It significantly increased the number of African American Ph.D. graduates in atmospheric and environmental sciences, and many of these graduates took jobs as NOAA scientists. The program also supported more than 150 research collaborations involving NOAA and minority-serving institutions.

NSF Research Experiences for Undergraduates Program. A 2006 evaluation carried out by SRI International examined NSF's REU and other undergraduate research programs (Russell et al., 2006). The effectiveness of the earth science REU program was not specifically examined. The evaluation used surveys of participants and recipients of bachelor's degrees to assess the characteristics of participants, why faculty and students choose to participate, and the impacts of different types of research experiences on students' academic and career decisions. The results showed that undergraduate research experiences increased participants' understanding of the research process and confidence in their ability to conduct research. The experiences also raised their awareness of STEM career options and informed their graduate school and career decisions. Among the report's recommendations was that evaluations could be strengthened by participant feedback on program strengths, weaknesses, and possible improvements.

NSF Opportunities for Enhancing Diversity in the Geosciences Program. The American Institutes for Research assessed the OEDG Program in 2010, based on their annual reviews of the impacts and rigor of evaluation activities of OEDG projects (Huntoon et al., 2010). The report identified successful OEDG projects as well as those that could not demonstrate success because of poor evaluation or data collection, and used these examples to develop best practices. Overall, the review found that the OEDG portfolio has produced an impressive array of successes in meeting OEDG Program goals, which are primarily aimed at exposing or involving underrepresented minorities in earth science. The report also made recommendations for improving data collection and evaluation of OEDG projects (e.g., requiring that proposals identify goals, outcomes, and an evaluation plan; documenting demographics of providers and participants; monitoring impacts).

EPA Science to Achieve Results Graduate Fellowship Program. An NRC review committee developed four metrics and gathered information, primarily surveys of former fellows, needed to evaluate them (NRC, 2003). The metrics focused on the selection process and outcomes (completion of a degree, publication of research, and a career in environmental science). The review found that the program's peer review process was effective in selecting high-quality fellows. Nearly all recipients completed their research and received a degree, and most had at least one peer-reviewed publication about their fellowship research. In addition, most were employed in an environmental science field. The report recommended that EPA collect information to quantify these results and better document the success of the program.

model) through intermediaries such as curriculum developers, exhibit designers, video producers, and print editors.

The goals of participants and providers differ for awareness activities, as do the outputs. In general, participants are looking for "fun" through positive interactions with peers and adults in novel contexts, while providers are looking to share research findings and the excitement of discovery or creation (Dierking et al., 2004). If an awareness activity is successful, participant outputs include enthusiasm and excitement for the positive interactions and some satisfaction for knowledge gained. The provider outputs include the number of individuals participating in the awareness activity and the participants' attitudes, intentions, and satisfaction with the activity.

Free-choice learning opportunities (i.e., those that take place outside the classroom) are a productive area for federal agencies to raise student awareness, but outcomes can be difficult to measure. Falk and Dierking (2000), for instance, noted that visitors to a museum exhibit often have difficulty expressing what they learned, unless they are asked to provide their own descriptions of the content of an exhibit. Furthermore, it is difficult, if not impossible, to randomly assign individuals to treatment groups (i.e., those receiving a specific intervention or experience) and control groups (i.e., those not receiving the specific intervention or experience) if they are voluntarily approaching a learning opportunity (Gaus and Mueller, 2011; Tucker et al., 2011). Observational, survey, or interview methods can return data useful for evaluation (Hein, 1998), but these methods are often expensive.

In the absence of adequate resources, the simplest method for evaluation is enumeration: counting participants or characteristics of participants. Enumeration data are useful for determining the scope and character of the participant pool, but they provide little information on how well an awareness program is working (Korn, 2012). To determine outcomes, the intentions of the program developers have to be aligned with the intentions of the participants through planned cycles of learning and practice. In such cycles, steps taken for planning, action, evaluation, and reflection are documented to show how results, drawn from evaluation data of different types, can be matched to the overall effort.

Another best practice is to carry out audience research. Researching the needs, interests, motivations, expectations, and learning styles of the intended audience enables the program design to be calibrated to the mission of an agency relative to the transaction (Seagram et al., 1993), in this case, raising awareness of earth science. Through audience research, agencies can generate evaluation data that match program content to the needs, interests, and capabilities of the intended audience (Kelly, 2004). Recruitment of participants is a critical and a constant activity, and provider organizations that share participant goals and accommodate group learning styles are among the most successful (Dierking et al., 2004).

Example Evaluations of Awareness Programs

Many of the earth science awareness programs discussed at the workshop employ enumeration of participants as the primary evaluation mechanism (e.g., NSF's Geoscience Education Program, USDA's Agriculture and Food Research Initiative programs). A few programs also make an effort to understand what participants have gained. For example, the National Park Service's (NPS's) National Fossil Day includes an online survey that allows participants to share what aspects of the program met their expectations and what they took away from the experience. Such efforts enable a closer alignment of the goals of the provider with the goals of the participants.

A comprehensive evaluation strategy is being employed by the Trail of Time project, an NSF–NPS–university collaboration not discussed at the workshop (Karlstrom et al., 2008). The project is aimed at helping visitors interpret Earth history along the south rim of the Grand Canyon. The project's evaluation plan includes both formative and summative evaluation, adjusting the content and design of exhibits based on participant learning outcomes. Although limited by sample size and potentially intrusive to the participant experience, the robust evaluation design allows providers to match content to participant motivations, capturing fine details of participant responses that would otherwise be lost.

Engagement

Engagement activities (e.g., earth science projects at science fairs, enrollment in an earth science major) provide opportunities for participants to develop their understanding of the Earth and the nature of earth science. Provider inputs to the logic model include specific content knowledge and skills as well as pedagogic expertise in designing engaging experiences. Outputs include participants' increased motivation to engage in learning activities beyond the formal science curriculum, increased understanding, and a more complete sense of ownership of a specific work product, project, or artifact through the application of new skills. Outputs for providers include the development of scientific habits of mind by participants, helping them to understand through participation in a professional community what it takes to become a scientist. Providers usually seek to enumerate participant characteristics, but they can also provide feedback that would further refine the interests of participants. Such feedback can be a critical incident that draws students into earth science.

The range and complexity of engagement activities present challenges to evaluation because short-term outputs may differ substantially from long-term outcomes. Nevertheless, common methods of assessment can generate useful data. The assessment systems used by state education agencies, for example, provide substantial data on the knowledge gained by students through formal instruction and some data on scientific skills. Positive feelings are commonly used as a proxy for assessing interest and motivation, but better indicators are available, including time on task; stored knowledge and value; responses to novelty, challenge, and complexity; and goal setting and self-regulation (Renninger, 2011). Evaluation models for experiential learning contexts (e.g., Fetterman and Bowman, 2002; Cachelin et al., 2009) can be used to assess knowledge, skills, and feelings. These approaches provide a strong basis for determining how to successfully engage students in earth science.

Example Evaluations of Earth Science Engagement Programs

Some federal engagement programs considered at the workshop specify outcomes focused on local, place-based needs (e.g., NSF's OEDG and Geoscience Teacher Training programs). Two programs use critical incident theory to understand how engagement opportunities influence subsequent academic and career choices. The NPS Geoscience-Teachers-in-Parks Program documents teacher feedback, the persistence of teacher's use of instructional materials, and student familiarity with the material to determine the importance of critical incidents in students' academic careers. Some projects in NSF's OEDG Program use critical incident theory to understand how and when students choose to engage in earth science and then pursue a career. The OEDG Program is a good example of a federal earth science education program that has been able to demonstrate success through a good evaluation strategy (Box 4.2).

An example of a successful engagement program not discussed at the workshop is the International Ocean Drilling Program's School of Rock, which is supported by NSF and uses data from ocean floor drill cores to document changes in the Earth system over time. A pilot evaluation of the ocean-going research experience was based on daily teacher connections journals, which recorded past experiences and knowledge, people, memorable events, instructional ideas, frustrations, and missed connections (St. John et al., 2009). A subsequent summative evaluation was based on interviews of teachers, who reflected on the efficacy of program implementation in their classrooms, and continued communication with participants. A 5-year follow-up (Collins et al., 2011), conducted through online surveys, included enumeration, an analysis of teacher lesson plans, and opportunities for professional development enabled by the experience. This evaluation identified critical elements of the program (e.g., teacher access to data and scientists) and acquired skills (e.g., knowledge transfer) and attitudes (e.g., science as a collaborative enterprise) through the material presented in classroom lessons.

Professional Preparation

Professional preparation opportunities (e.g., formal education, participation in professional society meetings, involvement in research, internships, postdoctoral fellowships) are aimed at a wide range of participants. High school students and undergraduates seek opportunities that provide a taste of the profession and help them acquire the knowledge and skills needed for an earth science career. Undergraduate and graduate students and new Ph.D. recipients seek opportunities that provide the full workplace experience or help them identify a suitable introductory position. These diverse audiences and objectives require a range of evaluation approaches. Approaches used in the two most common professional preparation activities—research experiences and internships—are described below.

Research Experiences

Among the reasons students get involved in undergraduate research are to experience what it is like to do science, to test their interest in an earth science career, or to develop specific job-related skills (e.g., Manduca, 1999). Providers of these experiences, namely researchers, seek to promote research activities, impart context-specific skills and scientific habits, and obtain results from specific learning goals. Inputs to the logic model include participants' interest and enthusiasm for "doing" science as well as providers' research interests and desire to mentor students as they enter the field. Outputs for undergraduate research projects include new knowledge and skills, increased persistence and interest in science careers, graduate school attendance, and higher graduation rates, especially among groups underrepresented in science (Thiry et al., 2011).

Calibrating the goals of undergraduate research with student expectations remains a significant evaluation challenge, although provider outcomes more consistent with participants' interests have been documented in supervisor evaluations of participants (Hunter et al., 2006). Relatively few empirical studies have examined whether students with undergraduate research experiences acquire higher order thinking skills in science (Kardash, 2000). Lopatto (2007) used the *Survey of Undergraduate Research Experience* (Lopatto, 2004) to investigate whether undergraduate research enhanced students' educational experience and attracted or retained them in science, technology, engineering, or mathematics (STEM) career paths. The surveys showed that the undergraduate research experience clarified or solidified students' graduate school plans. Participating students reported greater learning gains and a better overall undergraduate experience than nonparticipants. Students from underrepresented groups also showed greater retention rates than nonparticipating groups. These results were partly corroborated by Seymour et al. (2004), who found that participation in undergraduate research confirms students' prior career choices, increases their capacity to deal with ambiguity, and provides them with opportunities to take greater initiative for their own learning. Through a detailed review of the literature and a rigorous evaluation design, Thiry et al. (2011) found little evidence for the notion that participation in undergraduate research succeeds in recruiting students, attracting them to graduate school, or changing their choice of subjects. Thus, providers of professional preparation experiences may need to adjust their inputs into their logic model.

Research experiences are commonly evaluated by enumerating participation. For example, in a recent study, 85 percent of responding STEM graduates reported participating in some form of research experience (Thiry et al., 2011). However, a more effective approach is to match evaluation strategies to changes in experience format and duration. Research experiences range from projects with a research component that last a few weeks or a semester (Wagner et al., 2010; Gibson and Bruno, 2012) to fully immersive research experiences for undergraduates (e.g., Jarrett and Burnley, 2003; Gonzales-Espada and LaDue, 2006) to research at field stations and marine laboratories that last multiple semesters (Hodder, 2009). Efforts to define excellence in undergraduate research (e.g., research skills) and the logistics and infrastructure necessary to support high-quality work (e.g., Hensel, 2012) may help inform a comprehensive evaluation of undergraduate research experiences. Defining excellence requires both quantitative data (including enumeration of participants and their characteristics) and qualitative data (including surveys and interviews) and a careful matching of data to the goals of the program (Gonzalez-Espada and Zaras, 2006). Russell et al. (2006) concluded that there is no single way to define (and, by extension, to evaluate) the research experience, but that the sustained inculcation of enthusiasm for research provides the greatest impact.

Internships

Undergraduates seek internships to gain specific skills that will make them more competitive in the workplace, access to potential employers, and references to support their applications. For scientific internships, students seek broadly defined employment opportunities (Taylor, 1988) and the development of a scientific identity (Hsu et al., 2009). Providers, on the other hand, seek the successful completion of specific work products, the transfer of context-specific workplace skills, and access to a larger pool of suitable candidates for employment. Outputs include the acquisition of skills desired by the providers or themselves, a sense of ownership of the work product, and clarification of professional goals, even when the desired permanent job is not obtained. Providers gain work products at potentially lower costs, access to what they believe are top candidates for available positions, and satisfaction in providing a service to the profession. Reconciling the goals of the providers and the participants for evaluation purposes is aided by the transactional relationship between participants and providers.

The impact of internships on students and their hosts has been evaluated in a variety of ways, including interns' evaluations of their experiences, which provide useful feedback to the hosts (Morris and Haas, 1984; Girard, 1999), and supervisors' assessments of students' performance using the traditional academic grading structure (Cutting and Hall, 2008). Less available are clear evaluation findings that indicate whether the programs work or are cost-effective or whether interns gain knowledge, skills, and disposition in their chosen field (Schultz, 1981). The literature in science education (Schultz, 1981; Cutting and Hall, 2008; Hsu et al., 2009) and psychology (Shoenfelt et al., 2012) suggests that formative evaluation frameworks could be developed based on interactions between interns and their supervisors using an analysis of verbal transactions, work products, or written documentation. Key elements for summative evaluation include the appropriateness of the internship, provider and participant obligations and responsibilities, participant qualifications and expected competency gains, onsite supervision frameworks, and participant performance evaluation.

Example Evaluations of Earth Science Professional Preparation Programs

For the federal professional preparation programs discussed at the workshop, the most commonly employed evaluation strategy is the enumeration of participants. In addition to collecting enumeration data, the U.S. Geological Survey (USGS) Hydrologic Technician Internship Program, Youth Internship Program, and EdMap collect participant reports of satisfaction, provider evaluations of participants, and participant work products. The EdMap data show a relatively strong correlation of participant and provider responses on performance evaluations, onsite supervision frameworks, and obligations and responsibilities. However, the relationship between the expected competency gains and the appropriateness of internship opportunities is less clear. Adding an examination of work products and longitudinal tracking of participants as they move into the workforce would improve the evaluation with little added effort. Programs that collect these data include NOAA's Educational Partnership Program, which analyzes participant work products and tracks the transition of participants to the workforce, and NSF's Earth Sciences Postdoctoral Fellowships program, which collects some information on the workforce transition. Overall, providers that collect all of the information described (enumeration, self-reports, supervisor evaluations, work product analysis, and tracking) in a systematic, rigorous manner have a greater chance of aligning their goals and outputs with those of the participants.

SYSTEM-LEVEL EVALUATION

Evaluations at the various stages of the framework provide important information on how well an education and training program is achieving a goal of awareness, engagement, or professional development. Evaluations encompassing all activities in the framework could be used to find imbalances in effort and connections and gaps between activities at different stages of the framework. It could also provide a measure of the extent to which the portfolio of education and training programs offered by various organizations is changing earth science pathways.

In a system-level evaluation, the size and effectiveness of individual programs is viewed in the context of information about (a) levels of activity at various points along the path and (b) the status of the system objective. Broad indicators of program activities at various stages of the framework can be obtained by aggregating information from individual program evaluations. For example, the sum of earth science exhibits or classes and the number of people exposed to them can provide a measure of national awareness of earth science. Such measures can be supplemented with in-depth evaluations aimed at providing insight on the dynamics of the system at the various stages. Targeted program evaluations that measure activities and outcomes would increase understanding of how to create effective programs, and qualitative studies would show how individuals find the opportunities and what they learn from them.

As noted in Chapter 3, individuals travel different pathways to an earth science career, sometimes skipping stages or moving back and forth across stages of the system. A system-level evaluation would take account of the networks that help individuals find a path through the system. The presence, size, and interconnectedness of organizations in the various networks (e.g., university consortia, cultural and ethnic affinity organizations) can all be measured. Network analysis of the connections can be based on unobtrusive indicators such as Web links and common themes in public statements. Communication and dissemination efforts are particularly easy to measure, and they intersect with the awareness indicators described above.

SUMMARY AND CONCLUSIONS

Program evaluations provide a means for determining whether a program is succeeding and why. External evaluations have demonstrated successes in the OEDG, EPP, and STAR programs. The other federal programs considered in this report have not been evaluated and most were not designed to facilitate evaluation: some program goals are too broad to develop criteria for success; the goals and criteria do not always match; and the criteria and data collection emphasize what is easy to measure, not what the program is trying to achieve. These programs may be successful, but the data were too sparse and uneven in quality to make that determination. The difficulty of identifying successful programs (Task 3) and determining what made them successful (Task 4) underscores the importance of incorporating evaluation into program design.

Rigorous evaluation approaches commonly use a logic model to define who the program is trying to reach, what it is trying to achieve, what resources it requires (inputs), and how to translate program resources into near-term results (outputs) and long-term outcomes. Each program needs its own evaluation design and criteria for success. Enumeration, pre- and post-testing, observations of participants or providers, work product analysis, and determination of long-term plans and satisfaction with experiences are all useful tools for evaluation.

The framework of opportunities described in Chapter 3 can be used to conceptualize evaluation of individual programs and suites of programs with a collective goal of building earth science pathways to careers. Each stage of the framework (awareness, engagement, professional development) has its own input, activity, output, and outcome measures. Careful attention to input and activity measures would ensure that the goals of participants and providers are aligned. Measures across several fiscal years are commonly needed to assess long-term outcomes. Although more time-consuming and costly, long-term measures can demonstrate program impact as well as its sustainability.

A system-level evaluation, encompassing all activities within the framework or at a stage of the framework (e.g., engagement), could be used to identify imbalances in effort and gaps, enabling agencies to determine where future education and training efforts may be useful. Broad indicators of program activities could be developed by aggregating relevant information from individual program evaluations, and supplemented with targeted program evaluations aimed at understanding how to create effective programs. Network analysis of the programs in the system could reveal which connections among participating organizations help move individuals through the system, and qualitative studies would help show how individuals find education and training opportunities and what they learn from them.

5

Broadening the Participation of Underrepresented Groups

A key goal of federal government recruitment policies is to attain a workforce that draws from all segments of society and that leverages diversity to deliver the best public service (OPM, 2011). However, the federal earth science workforce—and the academic programs that produce graduates—does not yet mirror the ethnic, racial, and gender diversity of the U.S. population. For example, underrepresented minorities (African American, American Indian, and Hispanic or Latino of any race) composed 30 percent of the U.S. population in the 2010 Census, but received only 7.2 percent of earth science bachelor's degrees awarded in 2009 (NSB, 2012). Underrepresented minorities make up 3.5 percent of earth science-related positions at the U.S. Geological Survey (USGS; Box 5.1)[1] and between 2.2 and 8.1 percent of all geoscience and environmental science occupations (2003–2009 average; Gonzales and Keane, 2011). Women comprise 51 percent of the U.S. population and received 39 percent of bachelor's degrees in geoscience (NSF, 2013). They hold 21 percent of USGS earth science-related positions (Box 5.1) and 30 percent of all geoscience and environmental science occupations (Gonzales and Keane, 2011). This chapter describes the types of programs that have succeeded in attracting or retaining minorities and women on earth science pathways and the factors that made these programs successful, based on results published in the literature and discussions with experts on earth science diversity programs at the committee's workshop. These programs could help federal agencies leverage their earth education and training efforts to improve their recruitment of a diverse population in both high school and college (the committee's Task 6).

INCREASING THE PARTICIPATION OF UNDERREPRESENTED MINORITIES

Most federal agency guidelines and operational definitions of diversity related to program design and evaluation are focused on race/ethnicity, gender, physical (dis)ability, socioeconomic class, and increasingly on returning student or first-time-in-college status and veteran status. Under-

[1] Comparable figures from other federal agencies are not publicly available and may be higher or lower depending on factors such as the agency mission, mix of occupations, and proportion of federal employees and contractors.

BOX 5.1 Diversity of the Earth Science-Related Workforce at the U.S. Geological Survey

At the request of the committee, the USGS provided the current race/ethnicity and gender profile of its earth science-related occupations (i.e., geology, geophysics, hydrology, general physical science, physical science technicians, and hydrologic technicians), which make up 41 percent of its workforce. For these occupations, underrepresented minorities (African American, American Indian, and Hispanic or Latino) compose 3.5 percent of employees and all minorities (including Asian, Native Hawaiian or Other Pacific Islander, and two or more races) compose 6.2 percent of employees. Hispanic/Latino and Asian are the largest minority groups, each composing 2.8 percent of the workforce, followed by African American (1.0 percent), American Indian or Alaska Native (0.8 percent), two or more races (0.7 percent), and Native Hawaiian or Other Pacific Islander (0.2 percent). Diversity varies by occupation, with a relatively high fraction of underrepresented minorities in general physical science, physical science technician, and hydrologic technician positions. Asians are more represented in geology, hydrology, and geophysics positions than other minority groups.

Women currently compose only 21 percent of the USGS earth science-related workforce. Their representation is greatest in general physical science, physical science technician, and geology occupations.

SOURCE: Jo-Ann J. Dominique, Office of Diversity & Equal Opportunity, U.S. Geological Survey.

representation is also typically taken into account, using dimensions of identity easily accounted for by the U.S. Census Bureau. Although the 2010 Census began to account for an increasingly multiethnic and mixed-ethnicity population in the country, only Black, American Indian, and Hispanic or Latino of any race are considered "underrepresented minorities."

The National Research Council report *Expanding Underrepresented Minority Participation: America's Science and Technology Talent at the Crossroads* (NRC, 2011) laid out a roadmap for increasing the participation of underrepresented minorities in science, technology, engineering, and mathematics (STEM) education and for improving the quality of their education. The report found that of all possible actions that could be taken by academic institutions, government agencies, scientific societies, or industry, two would have immediate impact on critical transition points for underrepresented minorities: (1) undergraduate retention programs that increase graduation rates and (2) teacher preparation and student programs that increase participation in undergraduate and graduate education (NRC, 2011). Proven interventions for underrepresented minorities have been most thoroughly explored for academic education programs, but they are also applicable to federal agency education programs. Effective programs include the following:

- Summer programs that include or target middle school, high school, and undergraduate students (NRC, 2007, 2011). These programs stimulate interest through hands-on research and develop student cohorts that provide mutual support. In the context of the framework presented in Chapter 3, these are examples of awareness or engagement activities.
- Research experiences at the undergraduate and graduate levels (NRC, 2007, 2011). These programs encourage the further development of competence in and identification with the field. As discussed in the Chapter 3 framework, research experiences can help guide students to professional preparation.
- Opportunities for undergraduate and graduate students to network, participate in national conferences, present research results, and join study groups, social activities, tutoring, peer-to-peer

support, and mentoring programs (NRC, 2007, 2011). These professional preparation opportunities help socialize students within a discipline, promote academic success, and prepare them for careers. Mentors can play a key role in providing information, guidance, and support at critical decision points in students' careers.

- Financial programs that are based on need or are targeted at supporting undergraduate and graduate study (Smith, 1997; NRC, 2011). Affordability is key to the success of underrepresented minority students, and financial assistance is commonly required to provide access to adequate facilities, equipment, and course curricula.
- Efforts to lower barriers to the participation of underrepresented minorities in college, such as developing K–12 STEM outreach activities to cultivate potential future students; establishing strong connections between programs and institutions; and developing, implementing, and enforcing admissions policies that increase diversity of the student population (NRC, 2007, 2011). Such efforts address some of the system-level challenges described in Chapter 3.

The above programs open doors of opportunity to underrepresented minorities, but they could also help attract and retain students of all backgrounds. An example discussed at the workshop is GeoFORCE, a program established at the University of Texas in 2005 and aimed at bringing young people, particularly underrepresented minorities, into earth science.[2] The program engages more than 600 eighth-grade students in summer field trips that introduce earth science concepts and emphasize hands-on science. Student cohorts continue through high school, building a foundation of geology expertise and a community that is sustained through a support network of peer cohorts, their adult mentors, and college students who previously participated in the program. Students also receive resources and mentoring to help them prepare for college, apply for admission and financial aid, and find summer internships and jobs.

GeoFORCE recruits girls and boys in high-minority, economically disadvantaged regions. In inner-city Houston, the school population is 35 percent Black, 61 percent Hispanic, and 82 percent economically disadvantaged. Between 40 percent and 65 percent of ninth graders fail to complete high school. In the rural southwest, 90 percent of students are Hispanic and 78 percent are economically disadvantaged. Dropout rates (5–45 percent) are substantially lower than in Houston, but less than 15 percent of adults have college degrees. In contrast, all students in the GeoFORCE program have graduated from high school and 97 percent have entered college.[3] Two-thirds of the college students are STEM majors, including 12 percent in earth science. These figures are significantly higher than national averages.

IMPROVING THE SUCCESS OF DIVERSITY PROGRAMS

Research suggests that programs that have increased the number of minority students graduating in STEM fields commonly take a comprehensive approach that includes the integration of students into college academic and social systems, the development of knowledge and skills, and support, mentoring, monitoring, and advising (e.g., Tinto, 1987; Seymour and Hewitt, 1997; Maton et al., 2000; Matsui et al., 2003). Lessons learned from federal earth science education and training programs also suggest that certain factors are important for creating success. Factors for success discussed at the committee's workshop are summarized in Box 5.2 and factors important for the success of projects in National Science Foundation's (NSF's) Opportunities for Enhancing Diversity in the Geosciences (OEDG) Program are summarized in Box 5.3.

[2] See http://jsg.utexas.edu/geoforce.

[3] Information provided by Eleanour Snow, University of Texas, based on data from the National Center for Higher Education Management Systems: Information Center, http://www.higheredinfo.org/.

BOX 5.2 Workshop Discussions on Ways to Increase Diversity

Key points raised by individuals at the workshop included the following:

- Follow best practices by engaging early with students, developing cohorts of students, connecting people to places, emphasizing field and real-world experiences, including math and reading preparation, involving families and communities, or linking with service learning projects.
- Link successful model programs into pathways leading to careers.
- Help students envision themselves working in earth science careers, including those that are culturally and societally relevant.
- Use previous student cohorts to acclimate and motivate new students.
- Find leaders who champion the programs.
- Obtain a sustained commitment and access to resources (e.g., financial aid, mentors) within and across organizations.
- Increase focus on minority-serving institutions and on community colleges and pathways to 4-year colleges.
- Create partnerships with new communities and partnership coalitions to expand existing projects and build engaging and inclusive experiences.
- Evaluate programs to determine success, including tracking students to determine outcomes after they leave the program.

BOX 5.3 Lessons Learned From NSF's OEDG Program

A review of the National Science Foundation's Opportunities for Enhancing Diversity in the Geosciences Program (Huntoon et al., 2010) identified best practices for OEDG projects:

- Incorporating strong mentoring components.
- Including role models (e.g., faculty or graduate students) as leaders in the project.
- Considering the professional development of all involved with the project.
- Planning for sustainability (e.g., through institutionalization or the acquisition of research equipment).
- Broadening a project's reach through time by encouraging people or organizations not directly involved to take actions that contribute to attainment of the project's goals.
- Demonstrating relevance of earth science (culturally, personally, or professionally) to the target audience.
- Developing or strengthening institutional partnerships and personal connections.
- Communicating with multiple stakeholders throughout the project.
- Appreciating/accommodating the perspectives of the target audience during the project design stage and throughout the project.
- Involving students in research and professional events within the earth science community.
- Involving participants in field experiences.

Huntoon et al. (2010) also identified practices to avoid, including one-time or short interventions, intervening too late in students' careers, providing inadequate financial support, and providing insufficient advance training to investigators running the programs.

The workshop discussions and lessons learned from the OEDG review are consistent with published findings. In a special issue of the *Journal of Geoscience Education*, Riggs and Alexander (2007) found that programs that successfully support minority students in the earth sciences share a common set of factors. First and foremost, they have components that are deeply rooted in specific ethnic and cultural communities (e.g., Huntoon and Lane, 2007; Miller et al. 2007; Pride and Olsen, 2007; Riggs et al., 2007). Typically a few individuals in each program are intimately connected to the concerns, needs, and aspirations of a particular minority group and have also built bonds of trust and friendship in these communities. These connected academics and community members understand how to reach and engage potential earth science students.

Second, the universities and colleges involved in successful programs are often connected into networks and collaborations that leverage the strengths of each academic partner and provide a clear educational pathway, often from community college into graduate work (e.g., Gilligan et al., 2007; Pandya et al., 2007; Robinson et al., 2007). These networks are commonly supported by the upper administration of these universities and colleges, which provides not only recruitment support, but also access to programs that may be beyond the means of individual faculty or departments (e.g., Stokes et al., 2007). Third, funding, particularly from federal agencies, is critical to spur development of new programs or to augment existing ones (e.g., Chigbu et al., 2007; Pyrtle and Williamson-Whitney, 2007). Long-running programs may be supported by several organizations, including federal agencies, private companies, state agencies, and universities.

An example of long-running collaboration among academia, industry, and government discussed at the workshop is the Cooperative Developmental Energy Program,[4] which was established in 1983 and is hosted at Fort Valley State University, a historically Black university. The program currently operates as a 3+2 program—minority and women students major in biology, chemistry, or math at Fort Valley State University for the first 3 years, then transfer to a partner university for 2 years to complete degree requirements in an energy field, such as geology, geophysics, health physics (nuclear power industry), or engineering. Students receive full scholarships and internship opportunities supported by partner corporations (including oil/gas companies and electric power utilities), government agencies, and universities. To date, the program has graduated 27 geologists and geophysicists, 7 health physicists, and 76 engineers. About half have gone on to careers in the energy industry, and all graduates are employed in STEM fields.

INCREASING THE PARTICIPATION OF WOMEN IN EARTH SCIENCE EDUCATION

The participation of women in earth science education has been increasing for several decades, although women have not yet reached parity with men (Gonzales and Keane, 2011). Women currently receive 39 percent of bachelor's degrees, 47 percent of master's degrees, and 41 percent of doctorate degrees in geoscience (NSF, 2013). A range of social factors undermines the progress of women in earth science. During formal education, women often suffer from a lack of mentors and poor guidance and advice (Holmes and O'Connell, 2005, 2007). Montelone et al. (2006) argued that the way earth science departments market and represent their academic field on Web sites presents an image that is less inclusive for women and minorities by, for example, omitting images of diverse individuals and women in prominent roles. Studies from chemistry, physics, and engineering (e.g., Etzkowitz et al., 2000; Hodgson et al., 2000; Whitelegg et al., 2002; RSC, 2008)[5] suggest that the lack of family-friendly policies, challenges with mentors and advisors, and an often unsupportive

---

[4] See www.fvsu.edu/academics/cdep. Partner universities include Georgia Tech; Pennsylvania State University; University of Arkansas; University of Nevada, Las Vegas; University of Texas, Austin; and University of Texas, Pan American.

[5] See also meeting and workshop reports of the NRC Committee on Women in Science, Engineering, and Medicine at http://sites.nationalacademies.org/PGA/cwsem/index.htm.

and inflexible workplace culture contribute to the early departure of women from Ph.D. and academic tracks for industry or other fields.

The literature on barriers to the participation of women in earth science and how to overcome them is sparse and is commonly based on surveys and samples of convenience. The nature of earth science and pathways to earth science careers differ from those of other scientific fields, and so the literature on women in science in general is not always directly applicable to women in earth science. Consequently, it is not clear what types of education and training programs might be most effective for keeping women on earth science pathways. Moss-Racusin et al. (2012) suggest that addressing issues of unconscious biases may increase the representation of women in science.

LEVERAGING EDUCATION EFFORTS TO IMPROVE RECRUITMENT OF UNDERREPRESENTED GROUPS

Task 6 of the committee was to describe ways that federal agencies can leverage their earth education and training efforts to improve their recruitment of a diverse population in both high school and college. Agencies can strengthen the participation of minority students in earth science by supporting education and training programs that follow the effective practices discussed above. By working with other agencies and organizations, federal agencies could stretch the resources available for meeting their diversity goals.

To gather input on ways to leverage resources, the committee discussed the topic at the workshop (see Box 5.4) and also asked federal education program managers three questions:

1. What partnerships (federal, state, local government; academic; industry; nongovernmental organization; others) have you formed (whether active or inactive) and how have they contributed to the success of your earth science education project/program?
2. What other partners would it be constructive to work with in the future and why?
3. What are the barriers (if any) to working with other federal agencies to leverage earth science education programs, and how have you overcome them?

Only four agencies (USGS, NSF, Environmental Protection Agency, and National Oceanic and Atmospheric Administration) answered the questions. Their responses suggest that they partner primarily with other federal agencies working on community-wide initiatives (e.g., earth science literacy) or issues of mutual interest (e.g., the USGS and NSF fund summer interns through a UNAVCO program), with university diversity programs (e.g., GeoFORCE), and with professional societies, which carry out supporting activities (e.g., connecting agencies with university programs, conducting workshops at universities). A few agencies have leveraged federal dollars to gain contributions from private companies, mainly oil companies. Respondents desired increased collaboration, especially with industry and with professional societies that could connect agencies to private companies. Barriers to increased collaboration include different discipline foci, limited flexibility on funding mechanisms, poor familiarity with the various education and training programs, and a lack of time to seek collaborators or pursue partnerships. Some ideas for developing or strengthening collaboration are discussed below.

Interagency Collaboration

The agency responses to the committee's questions and the workshop discussions suggest that many federal earth science education and training programs operate in isolation. The framework described in Chapter 3 shows how agencies can place their programs in a common context, which would help them discover other programs with similar goals, identify program gaps and overlaps,

BOX 5.4 Workshop Discussions on Ways to Leverage Resources

Key points raised by individuals at the workshop included the following:

- Create a community of practitioners across federal agencies, professional societies (especially minority-serving societies), academic institutions, and industry to build a program framework and to share information and best practices from successful programs.
- Look for synergies in areas such as goals, missions, and data sets.
- Leverage existing programs (e.g., NSF's OEDG Program, University of Texas' GeoFORCE program, Fort Valley State University's Cooperative Development Energy Program) to create cohesive education and training networks.
- Partner with organizations that offer complementary strengths (e.g., expertise, geographic location, recruitment reach, convening capability) or different funding abilities.
- Divide work on joint projects such as program solicitations and outreach to broaden capacity.
- Share information tools, such as Web tools that match opportunities with students.
- Use interagency memorandums of understanding to break down barriers and facilitate coordination for internships, outreach and recruitment activities, and teacher professional development.
- Develop common evaluation frameworks and surveys to improve the efficiency of evaluation efforts.

identify potential collaborations or divisions of labor, and take advantage of lessons learned and effective practices to strengthen programs. As noted above, collaborations are especially important for moving minority cohorts through the various education opportunities toward a career. Once partners and programs are identified, agencies could use memorandums of understanding, intra-agency special interest groups, or other mechanisms to establish cooperative relationships.

Collaboration with Professional Societies and Nongovernmental Organizations

A number of professional societies and nongovernmental organizations already collaborate with federal agencies on specific programs. They can also play a role in helping advance students through the framework of federal earth science education and training opportunities. Currently, it can be difficult for students to find available opportunities because each agency advertises its own programs. As a consequence, slots for participants may go unfilled. Professional societies, which already advertise academic vacancies, could partner with federal agencies to advertise openings for internships and other student opportunities. Professional societies and nongovernmental organizations focused on diversity (e.g., National Association of Black Geoscientists, Institute for Broadening Participation) play an especially important role in connecting underrepresented minority students to available education and training opportunities. Providing an essential resource to future professionals would benefit the societies and organizations, and increasing the visibility of programs would benefit the agencies and students. Easy access to information about federal opportunities for education and training may also help retain students in earth science.

Collaboration with Private Companies

The earth science knowledge and skills sought by private companies overlap with those sought by federal agencies. Among the skills commonly sought by employers in a variety of disciplines are critical thinking, complex problem solving, the application of knowledge in real-world settings, and effective communication (Hart Research Associates, 2013). An earth science employer, ExxonMobil, is routinely looking for minority graduates who have a quantitative focus, are proficient in English, and are able to thrive in corporate culture (e.g., balance risk and restraint, task- and result-oriented, team players).[6] Many of these skills have been identified as important for the earth science workforce (see "Earth Science Knowledge and Skills Identified in NRC Workforce Reports" in Chapter 1).

To find the hands-on problem-solving skills they need, businesses may sponsor programs that engage students in activities that develop these skills or create internships that expose students to real-world work environments (Stephens and Richey, 2013). Partnering directly with these businesses can raise concerns about using private money to support the missions of federal agencies. However, establishing coalitions of partners from federal agencies, private companies, universities, and professional societies (e.g., as has been done in GeoFORCE) would avoid these concerns while stretching federal dollars and bringing a wide range of expertise to bear on training the next generation of earth scientists. Professional society meetings—which draw presenters, exhibitors, and recruiters from all sectors—provide a venue for interested organizations to connect. Building such coalitions is not easy and it takes time to build trust and establish common goals and approaches. However, such partnerships could both benefit the profession and help federal agencies meet their missions.

SUMMARY AND CONCLUSIONS

Women have made substantial gains in earth science over the past several decades, and now obtain 39 percent of bachelor's degrees. The sparse literature suggests that with attention to biases and mentoring, it may be possible to narrow or eliminate the degree gap between women and men. Compared with women, the gains of minorities in earth science education have been modest. Although it may be possible to apply some lessons learned from increasing the participation of women in earth science to the challenge of increasing minority participation, convolution of issues of culture and ethnicity with issues of gender adds complexity and uncertainty to potential solutions. Increasing the participation of minorities in earth science will require not only effective practices within individual programs but also attention to linkages between programs and system-scale inequities such as uneven access to mentors or financial resources.

Studies suggest that a variety of interventions are needed to increase the participation of underrepresented minorities in STEM fields, including (a) research experiences, either hands-on summer programs at the middle school to undergraduate level or research projects at the graduate and undergraduate level; (b) networking opportunities; (c) financial assistance to support undergraduate and graduate study; and (d) efforts to lower barriers to participation, such as developing K–12 STEM outreach activities to cultivate future students. Education and training programs that succeed in attracting and retaining minority students have some common elements, including intimate connections with communities and linkages with other programs to create clear educational pathways. The latter underscores the importance of thinking about education and training programs in the context of a system of opportunities that moves students from awareness to preparation for an earth science career.

Although some federal agencies work together to leverage education resources, many earth science education and training programs operate in isolation. By mapping their programs onto a common framework of education and training opportunities, agencies could identify potential partners or divisions of labor, as well as share effective practices for attracting and retaining minority students. Collaborations with professional societies focused on diversity could help connect

[6] Presentation by Mike Loudin, project manager, ExxonMobil Campus Project, on September 17, 2012.

minority students to education and training opportunities, providing students with another avenue of information about open positions. In addition, broad coalitions among federal agencies, private companies, universities, and professional societies would stretch federal resources and bring a wide range of expertise to bear on building earth science pathways for more diverse students.

References

AGI (American Geosciences Institute), 2012, Pulse of earth science: National status of K-12 earth science education, state by state, available at http://www.agiweb.org/education/statusreports/2007/states/.

Cachelin, A., K. Paisley, and A. Blanchard, 2009, Using the significant life experience framework to inform program evaluation: The Nature Conservancy's Wings & Water Wetlands Education Program, *Journal of Environmental Education*, **40**, 2-14.

Chigbu, P., A. Jearld, and M.J. Fogarty, 2007, National Oceanic and Atmospheric Administration and Jackson State University Program in Fish Stock Assessment, *Journal of Geoscience Education*, **55**, 541-549.

Collins, J., L.W. Peart, S.K. Cooper, K.K. St. John, R.M. Leckie, S.A. Hovan, and J.V. Firth, 2011, Assessing the impacts of the first five years of School of Rock on-board and on-shore professional development program for educators: Lessons for other teacher professional development programs, American Geophysical Union, 2011 Fall Meeting, abstract #ED33C-05.

Co-STEM (Committee on STEM Education), 2012, *Coordinating Federal Science, Technology, Engineering, and Mathematics (STEM) Education Investments: Progress Report*, National Science and Technology Council, Washington, D.C., 53 pp., available at http://www.whitehouse.gov/sites/default/files/microsites/ostp/nstc_federal_stem_education_coordination_report.pdf.

Cutting, R.H., and J.C. Hall, 2008, Requirements for a workable intern/practicum in the environmental sciences: Experience for careers and graduate school, *Journal of Geoscience Education*, **56**, 120-125.

Dierking, L.D., K.M. Ellenbogen, and J. Falk, 2004, In principle, in practice: Perspectives on a decade of museum learning research (1994-2004), *Science Education*, **88** (supplement 1), S1-S3.

Etzkowitz, H., C. Kemelgor, and B. Uzzi, 2000, *Athena Unbound: The Advancement of Women in Science and Engineering*, Cambridge University Press, Cambridge, UK, 282 pp.

Falk, J.H., and L.D. Dierking, 2000, *Learning from Museums: Visitor Experiences and the Making of Meaning*, AltaMira Press, Walnut Creek, CA, 272 pp.

Fetterman, D., and C. Bowman, 2002, Experiential education and empowerment evaluation: Mars Rover educational program case example, *Journal of Experiential Education*, **25**, 286-295.

Flanagan, J.C., 1954, The critical incident technique, *Psychological Bulletin*, **51**, 327-357.

Gaus, H., and C.E. Mueller, 2011, Evaluating free-choice climate education interventions applying propensity score matching, *Evaluation Review*, **35**, 673-722.

Gibson, B.A., and B.C. Bruno, 2012, The C-MORE Scholars Program: Motivations for an academic-year research experiences for undergraduates program, *Journal of College Science Teaching*, **41**, 12-18.

Gilligan, M.R., P.G. Verity, C.B. Cook, S.B. Cook, M.G. Booth, and M.E. Frischer, 2007, Building a diverse and innovative ocean workforce through collaboration and partnerships that integrate research and education: HBCUs and marine laboratories, *Journal of Geoscience Education*, **55**, 531-540.

Girard, T.C., 1999, Interns' perceptions of internships: A look at work, supervision and appraisals, *Journal of Cooperative Education*, **34**, 42-48.

Gonzales, L., and C. Keane, 2011, *Status of the Geoscience Workforce 2011*, American Geological Institute, Alexandria, VA, 171 pp.

Gonzalez-Espada, W., and D.S. LaDue, 2006, Evaluation of the impact of the NWC REU program compared with other undergraduate research experiences, *Journal of Geoscience Education*, **54**, 541-549.

Gonzalez-Espada, W.J., and D.S. Zaras, 2006, Evaluation of the impact of the NWC REU program compared with other undergraduate research experiences, in *15th Symposium on Education*, American Meteorological Society, Atlanta, 9 pp.

Hart Research Associates, 2013, *It Takes More Than a Major: Employer Priorities for College Learning and Student Success*. An Online Survey Among Employers Conducted on Behalf of the Association of American Colleges and Universities, 14 pp., available at http://www.aacu.org/leap/documents/2013_EmployerSurvey.pdf.

Hein, G.E., 1998, *Learning in the Museum*, Rutledge, New York, 203 pp.

Hensel, N., ed., 2012, *Characteristics of Excellence in Undergraduate Research*, Council on Undergraduate Research, Washington, D.C., 72 pp.

Hodder, J., 2009, What are undergraduates doing at biological field stations and marine laboratories? *Bioscience*, **59**, 666.

Hodgson, B., E. Scanlon, and E. Whitelegg, 2000, Barriers and constraints: Women physicists' perceptions of career progress, *Physics Education*, **35**, 454-459.

Hoisch, T.D., and J.I. Bowie, 2010, Assessing factors that influence the recruitment of majors from introductory geology classes at Northern Arizona University, *Journal of Geoscience Education*, **58**, 166-176.

Holmes, M.A., and S. O'Connell, 2005, *Where Are the Women Geoscience Professors?* NSF-AWG Foundation Workshop, September 25-27, 2003, Washington, D.C., 41 pp., available at http://eas.unl.edu/~mholmes/images/Where%20are%20the%20Women%20Geoscientists.pdf.

Holmes, M.A., and S. O'Connell, 2007, Leaks in the pipeline: Why do women remain curiously absent from the ranks of academia? *Nature*, **446**, 346.

Houlton, H.R., 2010, *Academic Provenance: Investigation of Pathways That Lead Students into the Geosciences*, Master's Thesis, Purdue University, West Lafayette, Indiana, 153 pp.

Hsu, P., W. Roth, and A. Mazumder, 2009, Natural pedagogical conversations in high school students' internship, *Journal of Research in Science Teaching*, **46**, 481-505.

Hunter, A.-B., S.L. Laursen, and E. Seymour, 2006, Becoming a scientist: The role of undergraduate research in students' cognitive, personal, and professional development, *Science Education*, **91**, 36-74.

Huntoon, J.E., and M.J. Lane, 2007, Diversity in the earth sciences and successful strategies for increasing diversity, *Journal of Geoscience Education*, **55**, 447-457.

Huntoon, J.E., R. Levine, and C. Martinez-Sussmann, 2010, *Opportunities for Enhancing Diversity in the Geosciences (OEDG) Expert Panel: OEDG Program Review and Recommendations (Draft)*, American Institutes for Research, Washington, D.C., 54 pp. plus appendixes.

IPCC (Intergovernmental Panel on Climate Change), 2012, *Managing the Risks of Extreme Events and Disasters to Advance Climate Change Adaptation: A Special Report of Working Groups I and II of the Intergovernmental Panel on Climate Change*, C.B. Field, V. Barros, T.F. Stocker, D. Qin, D.J. Dokken, K.L. Ebi, M.D. Mastrandrea, K.J. Mach, G.-K. Plattner, S.K. Allen, M. Tignor, and P.M. Midgley, eds., Cambridge University Press, Cambridge, UK, and New York, 582 pp.

Jarrett, O., and P. Burnley, 2003, Engagement in authentic geoscience research: Evaluation of research experiences of undergraduates and secondary teachers, *Journal of Geoscience Education*, **51**, 85-90.

Jolly, E., P.B. Campbell, and L.K. Perlman, 2004, *Engagement, Capacity and Continuity: A Trilogy for Student Success*, GE Foundation, 25 pp., available from http://www.smm.org/ecc/.

Kardash, C.M., 2000, Evaluation of an undergraduate research experience: Perceptions of undergraduate interns and their faculty mentors, *Journal of Educational Psychology*, **92**, 191-201.

Karlstrom, K., S. Semken, L. Crossey, D. Perry, E.D. Gyllenhaal, J. Dodick, M. Williams, J. Hellmich-Brayn, R. Crow, N. Watts, and C. Ault, 2008, Informal geoscience education on a grand scale: The Trail of Time exhibition, *Journal of Geoscience Education*, **56**, 354-361.

Kelly, L., 2004, Evaluation, research and communities of practice: Program evaluation in museums, *Archival Science*, **4**, 45-69.

Korn, R., 2012, Why counting attendees won't cut it for evaluation in the 21st century: Planning and evaluating informal science programs, in *Connecting People to Science*, J.B. Jensen, J.G. Manning, M.G. Gibbs, and D. Daou, eds., Astronomical Society of the Pacific, Vol. 457, San Francisco, pp. 9-16.

Lent, R., S. Brown, and G. Hackett, 1994, Toward a unifying social cognitive theory of career and academic interest, choice, and performance, *Journal of Vocational Behavior*, **45**, 79-122.

Lent, R., S. Brown, and G. Hackett, 2000, Contextual supports and barriers to career choice: A social cognitive analysis, *Journal of Counseling Psychology*, **47**, 36-49.

Lent, R.W., H.B. Sheu, D. Singley, J.A. Schmidt, L.C. Schmidt, and C.S. Gloster, 2008, Longitudinal relations of self-efficacy to outcome expectations, interests, and major choice goals in engineering students, *Journal of Vocational Behavior*, **73**, 328-335.

Levine, R., R. González, S. Cole, M. Fuhrman, and K.C. Le Floch, 2007, The geoscience pipeline: A conceptual framework, *Journal of Geoscience Education*, **55**, 458-468.

Lopatto, D., 2004, Survey of Undergraduate Research Experiences (SURE): First findings, *Cell Biology Education*, **3**, 270-277.

Lopatto, D., 2007, Undergraduate research experiences support science career decisions and active learning, *Life Sciences Education*, **6**, 297-306.

Maltese, A.V., and R.H. Tai, 2010, Eyeballs in the fridge: Sources of early interest in science, *International Journal of Science Education*, **32**, 669-685.

Maltese, A.V., and R.H. Tai, 2011, Pipeline persistence: Examining the association of educational experiences with earned degrees in STEM among U.S. students, *Science Education*, **95**, 877-907.

Manduca, C.A., 1999, The Keck Geology Consortium, in *12th Keck Symposium Volume*, Carleton College, Minnesota, pp. 1-4, available at http://keckgeology.org/12thsymposiumvolume.

Maton, K.I., F.A. Hrabowski III, and C.L. Schmitt, 2000, African American college students excelling in the sciences: College and postcollege outcomes in the Meyerhoff Scholars Program, *Journal of Research in Science Teaching*, **37**, 629-654.

Matsui, J., R. Liu, and C.M. Kane, 2003, Evaluating a science diversity program at UC Berkeley: More questions than answers, *Cell Biology Education*, **2**, 117-121.

McLaughlin, J.A., and G.B. Jordan, 1999, Logic models: A tool for telling your program's performance story, *Evaluation and Program Planning*, **22**, 65-72.

Miller, K.C., T. Carrick, C. Martínez-Sussmann, R. Levine, C.L. Andronicos, and R.P. Langford, 2007, Effectiveness of a summer experience for inspiring interest in earth science among Hispanic-American high school students, *Journal of Geoscience Education*, **55**, 596-603.

Montelone, B.A., R.A. Dyer, C. Burack, and S.E. Franks, 2006, Designing welcoming and inclusive STEM department websites, in *Women in Engineering Programs and Advocates Network (WEPAN) National Conference*, Pittsburgh, PA, http://advance.ksu.edu/file_download/8/wepan06b.pdf.

Moss-Racusin, C.A., J.F. Dovidio, V.L. Brescoll, M.J. Graham, and J. Handelsman, 2012, Science faculty's subtle gender biases favor male students, *Proceedings of the National Academy of Sciences*, **109**, 16,474-16,479.

Morris, S.B., and L.J. Haas, 1984, Evaluating undergraduate field placements: An empirical approach, *Teaching of Psychology*, **11**, 166-168.

NRC (National Research Council), 2003, *The Measure of STAR: Review of the U.S. Environmental Protection Agency's Science to Achieve Results (STAR) Research Grants Program*, The National Academies Press, Washington, D.C., 192 pp.

NRC, 2007, *Rising Above the Gathering Storm: Energizing and Employing America for a Brighter Economic Future*, The National Academies Press, Washington, D.C., 592 pp.

NRC, 2009, The program evaluation context, in *Evaluating Occupational Health and Safety Research Programs: Framework and Next Steps*, The National Academies Press, Washington, D.C., pp. 25-35.

NRC, 2010, *NOAA's Education Program: Review and Critique*, The National Academies Press, Washington, D.C., 184 pp.

NRC, 2011, *Expanding Underrepresented Minority Participation: America's Science and Technology Talent at the Crossroads*, The National Academies Press, Washington, D.C., 286 pp.

NRC, 2013a, *Emerging Workforce Trends in the U.S. Energy and Mining Industries: A Call to Action*, The National Academies Press, Washington, D.C., 390 pp.

NRC, 2013b, *Future U.S. Workforce for Geospatial Intelligence*, The National Academies Press, Washington, D.C., 172 pp.

NSB (National Science Board), 2012, *Science and Engineering Indicators 2012*, National Science Foundation, NSB 12-01, Arlington, VA, available at http://www.nsf.gov/statistics/seind12/.

NSF (National Science Foundation), 2013, *Women, Minorities, and Persons with Disabilities in Science and Engineering: 2013*, Special Report NSF 13-304, National Center for Science and Engineering Statistics, Arlington, VA, available at http://www.nsf.gov/statistics/wmpd/.

OPM (Office of Personnel Management), 2011, Government-wide diversity and inclusion strategic plan, Office of Diversity and Inclusion, 9 pp., available at http://www.opm.gov/diversityandinclusion/reports/GovernmentwideDIStrategicPlan.pdf.

Pandya, R.E., S. Henderson, R.A. Anthes, and R.M. Johnson, 2007, BEST practices for broadening participation in the earth sciences: Strategies from the UCAR Significant Opportunities in Atmospheric Research and Science (SOARS) Program, *Journal of Geoscience Education*, **55**, 500-506.

PCAST (President's Council of Advisors on Science and Technology), 2012, *Engage to Excel: Producing One Million Additional College Graduates with Degrees in Science, Technology, Engineering and Mathematics*, Report to the President, Executive Office of the President, February 2012, 103 pp., available at http://www.whitehouse.gov/sites/default/files/microsites/ostp/pcast-engage-to-excel-final_2-25-12.pdf.

Pride, C.J., and M.M. Olsen, 2007, Enhancing diversity in the earth sciences through providing HBCU science majors training in natural history interpretation and teaching experiences in predominantly African-American communities, *Journal of Geoscience Education*, **55**, 550-559.

Pyrtle, A.J., and V.A. Williamson-Whitney, 2007, The minorities striving and pursuing higher degrees of success (MS PhD's) in Earth system science professional development initiative: Enhancing intellectual merit and broader impact of diversity through alignment of vision, goals and objectives, and measurement, *Journal of Geoscience Education*, **55**, 514-521.

Renninger, K.A., 2011, On defining and assessing engagement, interest, curiosity, and motivation in informal science learning, Commissioned paper from Summit on Assessment of Informal and Afterschool Science Learning, Board on Science Education, National Research Council, June 11-12, 2012.

Riggs, E.M., and C.J. Alexander, 2007, Broadening participation in the earth sciences, *Journal of Geoscience Education*, **55**, 445-446.

Riggs, E.M., E. Robbins, and R. Darner, 2007, Sharing the land: Attracting Native American students to the earth sciences, *Journal of Geoscience Education*, **55**, 478-485.

Robinson, L., J. Rousseau, D. Mapp, V. Morris, and M. Laster, 2007, An educational partnership program with minority serving institutions: A framework for producing minority scientists in NOAA-related disciplines, *Journal of Geoscience Education*, **55**, 486-492.

RSC (Royal Society of Chemistry), 2008, The chemistry PhD: The impact on women's retention, Special commissioned report on survey results, available at http://www.rsc.org/images/womensretention_tcm18-139215.pdf.

Russell, S.H., M.P. Hancock, and J. McCullough, 2006, *Evaluation of NSF Support for Undergraduate Research Opportunities: Draft Synthesis Report*, SRI International, Menlo Park, CA, 29 pp.

Schultz, J.M., 1981, The salaried, credited internship in environmental studies, *Journal of Environmental Education*, **12**, 7-12.

Scriven, M., 1991, Beyond formative and summative evaluation, in *Evaluation and Education: At Quarter Century*, M.W. McLaughlin and D.C. Phillips, eds., National Society for the Study of Education, Chicago, pp. 19-64.

Seagram, B., L. Patten, and C. Lockett, 1993, Audience research and exhibit development: A framework, *Museum Management and Curatorship*, **12**, 29-41.

Seymour, E., and N.M. Hewitt, 1997, *Talking About Leaving: Why Undergraduates Leave the Sciences*, Westview Press, Boulder, CO, 429 pp.

Seymour, E., A. Hunter, S.L. Laursen, and T. DeAntoni, 2004, Establishing the benefits of research experiences for undergraduates in the sciences: First findings from a three-year study, *Science Education*, **88**, 493-534.

Shoenfelt, E.L., J.L. Kottke, and N.J. Stone, 2012, Master's and undergraduate industrial/organizational internships: Data-based recommendations for successful experiences, *Teaching of Psychology*, **39**, 100-106.

Smith, D.G., 1997, *Diversity Works: The Emerging Picture of How Students Benefit*, Association of American Colleges and Universities, Washington, D.C., 154 pp.

St. John, K., R.M. Leckie, S. Slough, L. Peart, M. Niemitz, and A. Klaus, 2009, The Integrated Ocean Drilling Program "School of Rock": Lessons learned from an ocean-going research expedition for earth and ocean science educators, in *Field Geology Education: Historical Perspectives and Modern Approaches*, S.J. Whitmeyer, D.W. Mogk, and E.J. Pyle, eds., Geological Society of America Special Paper 461, pp. 261-273.

Stephens, R., and M. Richey, 2013, A business view on U.S. education, *Science*, **340**, 313-314.

Stokes, P.J., G.S. Baker, J.P. Briner, and D.J. Dorsey, 2007, A multifaceted outreach model for enhancing diversity in the earth sciences in Buffalo, NY, *Journal of Geoscience Education*, **55**, 581-588.

Taylor, M.S., 1988, Effects of college internships on individual participants, *Journal of Applied Psychology*, **73**, 393-401.

Thiry, H., S.L. Laursen, and A. Hunter, 2011, What experiences help students become scientists? A comparative study of research and other sources of personal and professional gains for STEM undergraduates, *Journal of Higher Education*, **82**, 357-388.

Tinto, V., 1987, *Leaving College: Rethinking the Causes and Cures of Student Attrition*, 2nd Ed., University of Chicago Press, Chicago, IL, 312 pp.

Tucker, M., J. Bricker, and A. Huerta, 2011, An approach to measuring the impact and effectiveness of educational science exhibits, *Journal of Applied Communications*, **95**, 6-20.

Underwood, W.D., 2008, Geoscience education in the secondary school, *The Leading Edge*, **27**, 1330-1332.

Wagner, J.R., R.D. Bixler, E.R. Carraway, S.M. Moysey, L.R. Murdoch, M.R. Schlautman, and R.D. Warner, 2010, Institutionalizing undergraduate research for geology majors through creative inquiry experiences, *Journal of College Science Teaching*, **39**, 30-37.

Whitelegg, E., B. Hodgson, E. Scanlon, and C. Donovan, 2002, Young women's perceptions and experiences of becoming a research physicist, in *Proceedings of 12th International Conference of Women Engineers and Scientists*, July 27-31, 2002, Ottawa, Canada.

Windschitl, M., J. Thompson, and M. Braaten, 2008, Beyond the scientific method: Model-based inquiry as a new paradigm of preference for school science investigations, *Science Education*, **92**, 941-967.

Appendix A

Legislative Authorities

The first task of the committee was to summarize the legislative authority for science, technology, engineering, and mathematics (STEM) education and training granted to federal agencies with substantial programs in earth science. These agencies include the U.S. Geological Survey (USGS) and National Park Service, both covered under Department of the Interior (DOI) authorities, National Science Foundation (NSF), Department of Energy (DOE), National Aeronautics and Space Administration (NASA), National Oceanic and Atmospheric Administration (NOAA), Environmental Protection Agency (EPA), U.S. Department of Agriculture (USDA), and the Smithsonian Institution. The following legislative authorizations for federal agencies engaging in STEM education were summarized from *Coordinating Federal Science, Technology, Engineering, and Mathematics (STEM) Education Investments: Progress Report* (Co-STEM, 2012) and from information provided by Robert Ridky, USGS, acquired while he was serving as the DOI representative on the National Science and Technology Council's Education Subcommittee. It includes authorities both for general STEM initiatives and for earth science education initiatives, defined in this report as excluding oceanic, atmospheric, and space science.

DEPARTMENT OF THE INTERIOR

The Youth Conservation Corps Act of 1970 established a permanent program in the USDA and DOI for young adults to perform tasks on lands and waters administered by the two departments. The acts also include the authority to fund the costs of projects carried out on public lands by other qualified youth or conservation corps.

The Public Lands Corps Act of 1993 established a federal corps of young adults to work on conservation projects on federal, Indian, and Hawaiian homelands in exchange for living expenses and educational benefits.

The National Geologic Mapping Act of 1992 (Public Law 102-285 and subsequent reauthorizations) established the USGS National Cooperative Geologic Mapping Program, which includes an education component "(A) to develop the academic programs that teach earth-science students the fundamental principles of geologic mapping and field analysis; and (B) to provide for broad education in geologic mapping and field analysis through support of field teaching institutes."

NATIONAL SCIENCE FOUNDATION

National Science Foundation Act of 1950 (42 USC § 1862)—"The Foundation is authorized and directed to initiate and support . . . science education programs at all levels in the mathematical, physical, medical, biological, social, and other sciences, and to initiate and support . . . engineering education programs at all levels." It also authorizes NSF to promote and strengthen research and education in science and engineering, and awards "grants to associate-degree-granting colleges, and consortia thereof, to assist them in providing education in advanced-technology fields, and to improve the quality of their core education courses in science and mathematics." They are further authorized to award grants to establish mathematics and science education partnership programs with nonprofits and higher education institutions, with the goal of improving elementary and secondary mathematics and science instruction.

America COMPETES Reauthorization Act of 2010 (Public Law 111-358)—Section 505 states that the director shall collect, acquire, analyze, report, and disseminate statistical data on the condition and progress of STEM education. Section 508(c)(5) authorizes a partnership program for innovation that will "broaden the participation of all types of institutions of higher education in activities to meet STEM workforce needs and promote innovation and knowledge transfer." Sections 515, 516, and 527 provide grants for undergraduate internships that integrate private-sector and STEM coursework, for cyber-enabled learning for the STEM workforce, and for "research-based reforms in master's and doctoral level STEM education that emphasize preparation for diverse careers utilizing STEM degrees," respectively.

DEPARTMENT OF ENERGY

Department of Energy Science Education Enhancement Act of 1990 authorized the Secretary of DOE to "establish programs to enhance the quality of mathematics, science, and engineering education." This included research opportunities for underrepresented groups, high school students and teachers, and higher education; inner-city and rural partnerships; and museum-based programs. Further, the act established centers of excellence in STEM at high-need public secondary schools located in regions served by the national laboratories.

The Methane Hydrate Research and Development Act of 2000 directs DOE to "promote education and training in methane hydrate resource research and resource development through fellowships or other means for graduate education and training."

The Energy Policy Act of 2005 contains a number of provisions regarding education. Section 971 directs the Office of Science to support education and outreach activities in energy science-related fields. Section 622 authorizes the Nuclear Regulatory Commission to provide scholarships and a fellowship program and Section 651 directs the agency to establish research activities with Hispanic-serving institutions, historically Black colleges or universities, and Tribal colleges. Section 954 authorizes the support "of fundamental nuclear sciences, engineering, and health physics research through a nuclear engineering education and research program." Section 983 establishes a science and engineering education pilot program.

America COMPETES Act of 2007—Section 5003(b)(1) directed DOE to appoint a director of science, engineering, and mathematics education with "the principal responsibility for administering science, engineering, and mathematics education programs across all functions of the department." It also established grants for the creation or expansion of "public, statewide specialty secondary schools that provide comprehensive science and mathematics (including technology and engineering) education to improve the academic achievement of students in science and mathematics."

NATIONAL AERONAUTICS AND SPACE ADMINISTRATION

The National Aeronautics and Space Act of 1958 (Public Law 85-568)—Section 102(d) mandates that "aeronautical and space activities of the United States shall be conducted so as to contribute materially" to "the expansion of human knowledge of the Earth and of phenomena in the atmosphere and space."

NASA Authorization Act of 2005 (Public Law 109-155)—Section 612 directs NASA to "develop or expand programs to extend science and space educational outreach to rural communities and schools." Section 615 directs the NASA administrator to "strive to ensure equal access for minority and economically disadvantaged students to NASA's education programs." Section 616 authorizes NASA to award grants and cooperative agreements with museums and planetariums for the enhancement of programs "related to space exploration, aeronautics, space science, earth science, or microgravity."

NASA Authorization Act of 2008 (Public Law 110-422)—Section 703 states that "NASA's educational programs are important sources of inspiration and hands-on learning for the next generation of engineers and scientists which should be supported." Section 704(a) encourages NASA to include other federal agencies in its planning efforts to use the International Space Station (ISS) National Laboratory for STEM education activities. Section 704(c) directs NASA to "continue its emphasis on the importance of education to expand opportunities for Americans to understand and participate in NASA's aeronautics and space projects by supporting and enhancing science and engineering education, research, and public outreach efforts."

NASA Authorization Act of 2010 (Public Law 111-267)—Section 504(6) directs NASA to provide initial financial assistance to the organizations managing the ISS National Laboratory to enable it to initiate the "development and implementation of scientific outreach and education activities designed to ensure effective utilization of ISS research capabilities . . . and the development of educational programs . . . including student-focused research opportunities for conduct of research in ISS national laboratory facilities."

America COMPETES Reauthorization Act of 2010 (Public Law 111-358)—Section 202 directs that NASA "shall develop and maintain educational programs to: [c]arry out and support research based programs and activities designed to increase student interest and participation in STEM, including students from minority and underrepresented groups; [i]mprove public literacy in STEM; [e]mploy proven strategies and methods for improving student learning and teaching in STEM; [p]rovide curriculum support materials and other resources. . . [and] create and support opportunities for enhanced and ongoing professional development for teachers . . ." Section 204 states that "the ISS represents a valuable and unique national asset which can be utilized to increase educational opportunities and scientific and technological innovation," and directs that NASA "evaluate and, where possible, expand efforts to maximize NASA's contribution to interagency efforts to enhance [STEM] education capabilities . . ."

U.S. DEPARTMENT OF AGRICULTURE

Land Grant University Statutes (Morrill Acts; Equity in Education Land-Grant Status Act of 1994)—USDA supports cooperative research and postsecondary agricultural education programs, with state partners being the land grant universities. In addition to the original land grant institutions from the Morrill Acts (which includes 18 historically Black land grant colleges of agriculture), 31 Native American colleges gained land grant status in 1994.

Agricultural Research, Extension, and Education Reform Act of 1998—Section 406 authorizes the Secretary of Agriculture to "establish an integrated research, education, and extension competitive grant program to provide funding for integrated, multifunctional agricultural research, extension, and education activities."

Food, Conservation, and Energy Act of 2008 (Public Law 110-246)—Section 7406 amends section 2(b) of the Competitive, Special, and Facilities Research Grant Act to authorize the Secretary of Agriculture to establish the Agriculture and Food Research Initiative (AFRI), a competitive grant program to provide funding for fundamental and applied research, education, and extension to address food and agricultural sciences. Grant priorities include natural resources and environment, among others.

ENVIRONMENTAL PROTECTION AGENCY

The National Environmental Education Act of 1990 (Public Law 101-619) calls for the EPA Office of Environmental Education to (1) develop and support programs to improve understanding of the natural and built environment and the relationships between humans and their environment; (2) support development and dissemination of model curricula, educational materials, and training programs for elementary and secondary students and other interested groups; and (3) manage federal grant assistance provided to local education agencies, institutions of higher education, and other not-for-profit organizations. The act also establishes an environmental education and training program to train educational professionals in the development and delivery of environmental education and training programs and studies. The act calls for EPA to provide internships to post-secondary students and fellowships for in-service teachers with agencies of the federal government.

NATIONAL OCEANIC AND ATMOSPHERIC ADMINISTRATION

The National Sea Grant College Program Act (Public Law 107-299) encourages National Sea Grant Colleges to engage in research, education, and outreach programs.

Coastal Zone Management Act (Public Law 109-58) and Section 1461 of the act, establishing the National Estuarine Research Reserve System—The act requires NOAA to provide opportunities for public coastal and marine education and interpretation.

Omnibus Public Land Management Act of 2009 (Public Law 111-11); Title XII – Oceans—Subtitle A–, Ocean Exploration Act calls for NOAA to create a National Ocean Exploration program that includes education and outreach activities to improve public understanding of the ocean and coastal resources. *Subtitle B–, Ocean and Coastal Mapping Integration Act* calls for NOAA to create up to three joint coastal and mapping centers to provide graduate education and training in ocean and coastal mapping sciences. *Subtitle C–, Integrated Coastal and Ocean Observation System Act of 2009*, calls for NOAA to create a national ocean, coastal, and Great Lakes observing system that includes public outreach and education activities.

America COMPETES Act of 2007 (Public Law 110-69)—Section 4002 directs NOAA to "carry out and support research based programs and activities designed to increase student interest and participation in STEM." The section also calls for NOAA to "create and support opportunities for enhanced and ongoing professional development for teachers."

America COMPETES Reauthorization Act of 2010 (Public Law 111-358)—Section 302 states that the educational programs developed by NOAA shall be designed to increase student interest and participation in STEM, improve public literacy in STEM, employ methods for improving student learning and teaching in STEM, and provide curriculum support materials that can be integrated with comprehensive STEM education.

Presidential Memorandum: America's Great Outdoors, April, 16, 2010—The memorandum calls for NOAA, along with other agencies, to create opportunities for the public to engage in environmental conservation activities and engage in educational experiences in outdoor environments managed by the federal government.

Stewardship of the Ocean, Our Coasts, and the Great Lakes, July 19, 2010 (Executive Order 13547)—The Executive Order encourages NOAA to foster public understanding of the value of the ocean, coastal resources, and the Great Lakes.

SMITHSONIAN INSTITUTION

An Act to establish the "Smithsonian Institution," for the Increase and Diffusion of Knowledge Among Men (9 Stat. 102, 1846)—The act created the Smithsonian Institution as a trust instrumentality of the United States for the "increase and diffusion of knowledge among men." This organic act included the first functions, including creating research laboratories as well as a museum, observatory, and library.

Appendix B

Workshop Agenda

National Academies Keck Center
500 Fifth Street, NW, Washington, D.C.
September 17-18, 2012

Monday, September 17

8:00	Continental breakfast available in the meeting room	
8:30	Welcome and overview of the workshop	*Arthur Goldstein, Chair*
8:40	The academic sea in which earth science swims	*Geoffrey Feiss* *GSA Foundation*
9:00	Panel on what makes a successful education and training program	
	Creating a successful earth science training and education program	*John McLaughlin* *McLaughlin Associates*
9:20	Building the hiring base: Thoughts on ExxonMobil's support for geoscience outreach programs	*Mike Loudin* *ExxonMobil*
9:30	Lessons learned from experience in starting and evaluating university programs: What works, what doesn't, and why	*Timothy Bralower* *Pennsylvania State University*
	Discussion (all morning presentations)	*All*
10:00	USGS motivations for the study and workshop	*Marcia McNutt* *USGS*
10:10	Break	

10:20	Charge to the working groups: Identify criteria for evaluating the success of earth science education and training programs	*Arthur Goldstein*
	Divide into four working groups	
12:00	Working lunch	
1:00	Working groups report back	
	Working group 1 Working group 2 Working group 3 Working group 4	*Lisa White or Raquel Gonzalez* *Isaac Crumbly or Matthew Dawson* *Eric Riggs or Jim Kahler* *Lisa Lauxman or Marcia Barton*
	Discussion	*All*
1:45	Flagship successful programs	
	Cooperative Summer Field Training Program	*Robert Ridky, USGS*
	Earth Science Research Experience for Undergraduates	*Lina Patino, NSF*
	Education Partnership Program	*Louisa Koch, NOAA*
	GeoFORCE Texas	*Eleanour Snow, University of Texas*
2:45	Charge to the working groups: Factors that make education programs successful	*Arthur Goldstein*
	Divide into four working groups	
4:15	Break and working groups prepare reports	
4:30	Working groups report back	
	Working group 1 Working group 2 Working group 3 Working group 4	*Pranoti Asher or Stephanie Stockman* *Robert Ridky or Louie Tupas* *Eric Jolly or Heather Houlton* *Lina Patino or Brandon Jones*
	Discussion	*All*
5:15	Plans for the next day	*Arthur Goldstein*

Tuesday, September 18

8:00	Continental breakfast available in the meeting room	
8:30	Overview of the second day of the workshop	*Arthur Goldstein*
8:35	Expanding participation of underrepresented groups	*Eric Jolly* *Science Museum of Minnesota*
9:05	Charge to the working groups: Identify successes and gaps in expanding the participation of underrepresented groups	*Arthur Goldstein*
	Divide into four working groups	
11:00	Break and working groups prepare reports	
11:20	Working groups report back	
	Working group 1	*Geoffrey Feiss or John Baek*
	Working group 2	*Timothy Bralower or Peter Lea*
	Working group 3	*Cathryn Manduca or Peter Lyttle*
	Working group 4	*Karl Turekian or Jill Karsten*
	Discussion	*All*
12:00	Working lunch	
1:00	Charge to the working groups: Identify program synergies and describe ways that federal agencies can leverage their education and training efforts to improve to recruitment of a diverse population	*Arthur Goldstein*
	Dividc into four working groups	
3:00	Break and working groups prepare reports	
3:20	Working groups report back	
	Working group 1	*Susan Cozzens or Don Sweet*
	Working group 2	*Elizabeth Day-Miller or Vince Santucci*
	Working group 3	*Eric Pyle or Lynne Murdock*
	Working group 4	*Louisa Koch or Eleanour Snow*
	Discussion	*All*
4:00	Synthesis and discussion Next steps for the committee	*Committee*
4:30	Workshop adjourns	

Appendix C

Workshop Participants

Pranoti Asher, American Geophysical Union
John Baek, National Oceanic and Atmospheric Administration
Marcia Barton, National Science Foundation
Timothy Bralower, Pennsylvania State University
Susan Cozzens, Georgia Institute of Technology
Isaac Crumbly, Fort Valley State University
Matthew Dawson, Geological Society of America
Elizabeth Day-Miller, Bridgewater Education Consulting
Elizabeth Eide, National Research Council
Michael Feder, White House Office of Science and Technology Policy
Geoffrey Feiss (emeritus), College of William and Mary
Deborah Glickson, National Research Council
Arthur Goldstein, Bridgewater State University
Raquel Gonzalez, American Institutes for Research
Linda Gundersen, U.S. Geological Survey
Heather Houlton, American Geosciences Institute
Eric Jolly, Science Museum of Minnesota
Brandon Jones, Environmental Protection Agency
Jim Kahler, U.S. Department of Agriculture
Jill Karsten, National Science Foundation
Louisa Koch, National Oceanic and Atmospheric Administration
Lisa Lauxman, U.S. Department of Agriculture
Peter Lea, National Science Foundation
Anne Linn, National Research Council
Mike Loudin, ExxonMobil
Peter Lyttle, U.S. Geological Survey
Cathryn Manduca, Carleton College
John McLaughlin, McLaughlin Associates

Marcia McNutt, U.S. Geological Survey
Lynne Murdock, National Park Service
Lina Patino, National Science Foundation
Eric Pyle, James Madison University
Robert Ridky, U.S. Geological Survey
Eric Riggs, Texas A&M University
Vince Santucci, National Park Service
Sally Goetz Shuler, Smithsonian Institution
Eleanour Snow, University of Texas, Austin
Stephanie Stockman, National Aeronautics and Space Administration
Don Sweet, U.S. Geological Survey
Louie Tupas, U.S. Department of Agriculture
Karl Turekian (emeritus), Yale University
Lisa White, University of California Museum of Paleontology
Nicholas Woodward, Department of Energy

Appendix D

Program Evaluation Information Provided by the Agencies

Program	Goals	Evaluation	Criteria for Success	Successes
USGS National Cooperative Geologic Mapping Program (EdMap)	• Train students in geologic mapping • Help fund academic research • Prepare students for real-world careers in the geosciences • Create collaborations between academic institutions, state geological surveys, and the USGS • Contribute to national geologic mapping efforts	• Survey of participants and providers	• Number of students educated in geologic mapping techniques	• Program benefits participants
USGS/NAGT Cooperative Summer Field Training Program	• Provide an opportunity to partner a highly able intern with a quality science mentor to work on a meaningful research project • Present research results at professional meetings or in refereed publications	• Surveys of participants and providers	• Number of participants • Success of participants in earth science careers	• Excellent interns are placed in research units • Program quality is recognized by USGS and earth science academic community
USGS Youth Internship Program	• Stimulate ongoing interest in science among college undergraduates • Build a pool of well-prepared new college graduates who could fill vacancies in the USGS	• Reports by participants and providers	• Fraction of participants that obtained permanent jobs • Program satisfaction	• Most available slots are filled • Many interns accept positions at USGS
USGS Hydrologic Technician Internship Program	• Stimulate ongoing interest in water science among college undergraduates • Build a pool of well-prepared new college graduates who could fill vacancies in the USGS	• Reports by participants and providers	• Fraction of participants that obtained permanent jobs • Program satisfaction	• Most available slots are filled • Many interns accept positions at USGS

NSF Earth Sciences Research Experience for Undergraduates (REU) Program	• Support research participation by undergraduate students	• External, formal evaluation in 2006 • Reports by providers	• Number of applicants to REU sites • Number of former participants in an earth science career	• Interest of undergraduate students
NSF Opportunities for Enhancing Diversity in the Geosciences (OEDG) Program	• Increase participation in the geosciences by African Americans, Hispanic Americans, Native Americans, Native Pacific Islanders, and persons with disabilities • Increase the perceived relevance of the geosciences among broad and diverse segments of the population	• External evaluation by an expert consulting firm using a logic model • External reviews every 3 years by a visiting committee • Reports by providers	• None specified	• New degree programs at minority-serving institutions (MSIs) • New partnerships between MSIs and 4-year and research universities • Better understanding how to recruit and retain minorities in earth science • Slight increase in number of undergraduate majors
NSF Earth Sciences Postdoctoral Fellowships	• Recognize beginning investigators of significant potential • Provide these individuals with research and education experience that will establish them in positions of leadership in the scientific community	• External reviews every 3 years by a visiting committee • Reports by participants on research advances • Information on career development of former fellows	• None stated	• A high fraction of participants are in tenure-track positions • Participants are active scientifically • Some participants have become mentors

continued

Program	Goals	Evaluation	Criteria for Success	Successes
NSF Geoscience Education (GeoEd) Program	• Improve the quality and effectiveness of formal and informal geoscience education at all educational levels • Increase the number of students pursuing geoscience education and career paths • Broaden participation of traditionally underrepresented groups in the geosciences • Promote public engagement in Earth system science	• External reviews every 3 years by a visiting committee • Reports by providers	• Widespread use of models for understanding geoscience content and strategies for developing spatial and systems-thinking skills • Number of peer-reviewed publications • Continued research support	• Incubation of the Digital Library for Earth System Education • Incorporation of big ideas that all citizens should know into the next-generation science standards • Creation of geoscience educators who use best practices from STEM education research to strengthen formal and informal education
NSF/NOAA/NASA Global Learning and Observations to Benefit the Environment (GLOBE) Program	• Improve student understanding of environmental and Earth system science across the curriculum • Contribute to scientific understanding of Earth as a system • Build and sustain a global community of students, teachers, scientists, and citizens • Engage the next generation of scientists and global citizens in activities to benefit the environment	• Internal evaluation (not specified)	• None specified	• A large number of K–16 students are engaged worldwide • A large community of teachers with improved pedagogy and content knowledge
NSF Geoscience Teacher Training (GEO-Teach) Program	• Improve the quality of geoscience instruction, primarily at middle and high school levels	ESSEA project: • Internal evaluation based on data from course site, participant surveys and products, and provider surveys and products	ESSEA project: • Resources developed and how much they have been used • Effect of courses on learner's knowledge of Earth system science and how to	ESSEA project: • Institutionalization of effective professional development and training methods over many universities • Funding to create more modules

			teach it • Effect of the program on faculty and their institutions	
		TESSE project: • External evaluation planned based on participant surveys, focus groups, phone interviews, observation of activities, and standardized teacher assessment	TESSE project: • Extent to which project goals, activities, and outcomes are understood by providers and participants • Extent to which program activities met stated goals • Extent to which teacher opportunities improved their understanding of earth science and ability to teach it • Success of processes and structures to improve collaboration	TESSE project: • Identification of challenges in extending a professional development model from research institutions to HBCUs
DOE Office of Science Graduate Fellowship (SCGF) Program	• Encourage the development of the next generation of scientific and technical talent in the United States who will pursue careers in research critical to the Office of Science mission	• Program goals, outputs, and outcomes defined in a logic model • Reports by providers and participants • Longitudinal tracking of participants • External review of WDTS portfolio in 2010 • External reviews every 3 years by a visiting committee	Long-term outcomes include • Pursuit of a career in DOE Office of Science-related research in academia, a DOE laboratory, or industry	• Program is too young to determine. The first cohort is nearing the end of the 3-year support period

continued

Program	Goals	Evaluation	Criteria for Success	Successes
DOE Summer of Applied Geophysical Experience (SAGE) Program	• Introduce students in geophysics and related fields to "hands-on" geophysical exploration and research • Apply research and geophysical exploration methods to solve specific problems	• Reports by participants and providers	• Rate of continuation to graduate school and careers in geophysics • Support and feedback from industry participants and visitors • Presentations at professional meetings and publications	• None specified
DOE Science Undergraduate Laboratory Internships (SULI) Program	• Encourage undergraduate students to pursue STEM careers	• Program goals, outputs, and outcomes defined in a logic model • Pre- and postparticipation surveys of participants • Longitudinal tracking of participants • External review of DOE laboratories' execution of the program • External review of WDTS portfolio in 2010 • External reviews every 3 years by a visiting committee	Near-term outcomes include • Completion of STEM degree • Pursuit of advanced degree in STEM • Preparedness for STEM career • Knowledge of DOE mission science and technology Long-term outcomes include • Pursuit of a career in DOE Office of Science-related research in academia, a DOE laboratory, or industry	• More than 90 percent of participants report that the internship helped prepare them for a STEM career • More than three-quarters of participants (2003-2011) are pursuing a career in STEM
DOE Community College Internship (CCI) Program	• Encourage community college students to enter technical careers relevant to the DOE mission	• Program goals, outputs, and outcomes defined in a logic model • Pre- and postparticipation surveys of participants • Longitudinal tracking of participants • External review of DOE laboratories' execution of the program	Near-term outcomes include • Completion of STEM degree • Pursuit of certification in STEM • Preparedness for STEM career • Knowledge of DOE mission science and technology	• None specified

		• External review of WDTS portfolio in 2010 • External reviews every 3 years by a visiting committee	Long-term outcomes include • Pursuit of a career in DOE Office of Science-related research in academia, a DOE laboratory, or industry	
NASA Gravity Recovery and Climate Experiment (GRACE) programs	None specified. Proposed outcomes are: • Conduct GRACE activities with 500 teachers with priority given to educators from underrepresented areas or with underrepresented audiences • Increase the knowledge of 500 secondary teachers about the Earth's history, Earth system science, and global climate change • Increase the knowledge of 500 students about earth science and global climate change • Make available selected activities of "changing mass = changing earth" programs through the University of Texas Center for Space Research Web site • Incorporate NASA satellite data and resources into the activities	• Reports by providers • External evaluations are planned	• Number of teacher and student participants • Field trip opportunities for students • Increase in knowledge of global climate change • Student interest in STEM careers	• None specified

continued

Program	Goals	Evaluation	Criteria for Success	Successes
USDA 4-H Environmental Education/Earth Science programs	• Increase awareness of science among youth • Improve science skills and knowledge among youth • Increase awareness of opportunities to contribute to society using science skills • Increase life skills among youth	• Internal evaluation (not specified)	• None specified	• 4-H enrollment • Professional development has positive impact on performance and satisfaction • High-quality programs • Creation of a new research database • Participants excel beyond their peers
USDA Agriculture and Food Research Initiative (AFRI) projects	• Increase technical competency in priority area(s) to ensure that the United States remains globally competitive in the knowledge age	• None yet	• Number of participants	• Participants complete their degree
USDA National Institute of Food and Agriculture Fellowships Grant Program	• Strengthen the ability of the nation's scientific community to meet the current and future challenges facing agriculture, forestry and food systems • Develop the technical and academic competence of doctoral candidates • Develop the research independence and teaching credentials of postdoctoral scientists	• Reports by providers	• None specified (project dependent)	• None specified (project dependent)
EPA Greater Research Opportunities (GRO) Undergraduate Fellowship Program	• Support undergraduate study in environmentally related fields	• External review by a visiting committee in 2009 • Reports by participants	• Number of awards • Number of completed degrees • Topic area distribution • Professional success of former fellows	• None specified

EPA Science to Achieve Results (STAR) Graduate Fellowship Program	• Defray the cost of study leading to advanced degrees in environmental science fields	• External review by a visiting committee in 2009 • Reports by participants • NRC review of STAR in 2003	• Number of awards • Number of completed degrees • Topic area distribution • Professional success of former fellows	• None specified
NOAA Educational Partnership Program (EPP) With Minority Serving Institutions	Cooperative Science Centers: • Increase the number of educated, trained, and graduated students, particularly from underrepresented communities in STEM fields that directly support NOAA's mission • Increase collaborative research efforts between NOAA scientists and researchers at NOAA EPP cooperative science centers	Cooperative Science Centers: • External review of cooperative science centers in their third year • NRC review of education program in 2010 • Surveys of providers and participants	Cooperative Science Centers: • Annual number of students from underrepresented communities who are trained and graduate in NOAA mission sciences • Annual number of students who are trained and graduate in NOAA mission sciences • Number of students completing experiential opportunities at NOAA facilities • Number of students who are hired by NOAA; NOAA contractors; other environmental, natural resource, and science agencies; academia; and the private sector • Number of collaborative research projects undertaken between NOAA and EPP cooperative science centers in support of NOAA operations • Number of students and faculty who participate in and complete postdoctoral-level research programs in support of the NOAA mission	Cooperative Science Centers: • None specified

continued

Program	Goals	Evaluation	Criteria for Success	Successes
			• Number of peer-reviewed papers published in NOAA mission sciences by faculty, postdoctoral fellows, and students sponsored by NOAA EPP • Funds leveraged with NOAA EPP funds • Number of participants engaged in NOAA mission-relevant learning opportunities	
	Graduate Sciences Program: • Increase the number of educated, trained, and graduated students, particularly from underrepresented communities in STEM fields that directly support NOAA's mission	Graduate Sciences Program: • NRC review of education program in 2010 • Surveys of providers and participants	Graduate Sciences Program: • Number of graduate sciences program students hired by NOAA	Graduate Sciences Program: • None specified
	Undergraduate Scholarship Program: • Increase the number of educated, trained, and graduated students, particularly from underrepresented communities in STEM fields that directly support NOAA's mission	Undergraduate Scholarship Program: • NRC review of education program in 2010 • Surveys of providers and participants	Undergraduate Scholarship Program: • Number of undergraduate scholarship students who are educated, trained, and graduate in NOAA mission sciences • Number of undergraduate scholarship students who pursue graduate work in NOAA mission sciences	Undergraduate Scholarship Program: • None specified

			• Number of undergraduate scholarship students who are hired by NOAA, NOAA contractors, and other natural resources and science agencies	
NPS/GSA Geoscientists-in-the-Parks Program	• Provide on-the-job earth science training for America's youth • Build technical capacity for parks • Enhance the public's understanding of earth science	• Reports by providers	• Completion of unmet earth science projects in NPS units • Communication of earth science to the public and park staff who are not geologists • Program growth	• Program has grown from 20 positions per year in 2007 to more than 100 positions per year in 2012
NPS/NAGT Geoscience-Teachers-in-Parks Program	• Provide cooperative exchange of learning and scientific research between the park, local earth science teachers, and communities • Advance educational and interpretive opportunities at the park • Develop a lifelong network with local communities, schools, and the park	• Participant feedback	• None specified	• Teachers gain career enhancement opportunities • Program increases the recognition of NAGT as a leader in promoting earth science education • Teachers sign up as park volunteers and return to the park after completing their internship • Products developed during the internship can be used by park staff for other educational purposes • Teachers continue using the park as an educational tool or destination • Students become more familiar with the park as an educational resource and more aware of the need to protect park resources • Teachers and students become stewards of parks and the environment

continued

Program	Goals	Evaluation	Criteria for Success	Successes
NPS National Fossil Day	• Promote public awareness and stewardship of fossils • Foster a greater appreciation of the scientific and educational value of fossils	• Survey of participants	• None specified	• Rapid expansion of partners embracing the mission • Positive feedback from partners and the media
Smithsonian Leadership and Assistance for Science Education Reform (LASER) Program	• Improve K–12 science education	• External reviews of the program and its components are planned	• Increased student achievement	• Statewide implementation of program (e.g., Washington state LASER)

NOTE: DOE = Department of Energy; EPA = Environmental Protection Agency; ESSEA = Earth System Science Education Alliance; GSA = Geological Society of America; HBCUs = historically Black colleges and universities; MSI = minority-serving institution; NAGT = National Association of Geoscience Teachers; NASA = National Aeronautics and Space Administration; NOAA = National Oceanic and Atmospheric Administration; NPS = National Park Service; NSF = National Science Foundation; STEM = science, technology, engineering, and mathematics; TESSE = Transforming Earth Systems Science Education; USDA = U.S. Department of Agriculture; USGS = U.S. Geological Survey; WDTS = Office of Workforce Development for Teachers and Scientists.

Appendix E

Biographical Sketches of Committee Members

Arthur Goldstein, *chair*, is the founding dean of the College of Science and Mathematics at Bridgewater State University, Massachusetts, a position he has held since August 2010. Prior to joining Bridgewater State University, he held appointments as a dean at the University of New England and the director of the Division of Earth Sciences at the National Science Foundation (NSF). While at NSF, Dr. Goldstein was involved in developing GeoTeach, a program aimed at improving the development of preservice and in-service secondary school teachers. Prior to his appointment at NSF, he was a professor of geology at Colgate University and served as department chair for 5 years. Dr. Goldstein was the co-chair of the NRC study on Scientific Ocean Drilling: Accomplishments and Challenges. He received a B.S. in geology from Kent State University and an M.S. and Ph.D. from the University of Massachusetts, Amherst.

Pranoti Asher is the education and public outreach manager for the American Geophysical Union (AGU), a professional society of the earth and space sciences with more than 60,000 members. Prior to joining AGU, she spent 18 years as an earth science faculty member at universities and community colleges. Dr. Asher is deeply interested in education, outreach, and workforce development for the earth sciences and is currently working on an NSF-funded study aimed at the development of earth and space science faculty and their students at 2-year colleges. She is also knowledgeable about education and outreach activities of other professional geoscience societies. She received her B.Sc. and M.Sc. in geology from the University of Bombay, India, and a Ph.D. in the geological sciences from the University of Connecticut.

Susan E. Cozzens is vice provost for Graduate Education and Faculty Affairs at the Georgia Institute of Technology, as well as a professor of public policy and director of the Technology Policy and Assessment Center. Her research interests are in science, technology, and innovation policies, with an emphasis on issues of equity, equality, and development. Dr. Cozzens is active in developing science and technology indicators as well as methods for assessing research. She is a fellow of the American Association for the Advancement of Science (AAAS), and past chair of the AAAS Committee on Science, Engineering, and Public Policy. Dr. Cozzens has served on nine previous

NRC committees, mostly related to program evaluation in a wide range of fields. She has a B.S. in sociology from Michigan State University and a Ph.D. in sociology from Columbia University.

Cathryn A. Manduca is director of the Science Education Resource Center at Carleton College. She is involved in a variety of projects that support improvements in undergraduate earth science education. Her areas of interest include bringing research results on teaching and learning into broader use in the earth sciences, understanding earth science expertise, and building strong earth science departments. She is the executive director of the National Association of Geoscience Teachers (NAGT) and serves on the AGU Outreach Committee and the American Institute of Physics Education Committee. Dr. Manduca is the lead investigator of a $10 million NSF grant to improve earth science education and to integrate the earth sciences across other academic disciplines. She received a B.A. in geology from Williams College and an M.S. and Ph.D. in geology from the California Institute of Technology.

Eric J. Pyle is a professor of geology in the Department of Geology and Environmental Science at James Madison University (JMU). He has a strong background in K–12 and college-level STEM education and program assessment and evaluation. Dr. Pyle was a member of the Earth and Space Science Design Team for the NRC's "A Curriculum Framework for K–12 Science Education." He is a past president of both the Virginia Association of Science Teachers and the West Virginia Science Teachers Association. He currently serves as co-director for the JMU Center for STEM Education and Outreach, coordinator for Science Teacher Preparation in the College of Science and Mathematics at JMU, and as an NAGT counselor for Virginia. He has a B.S. in earth science from the University of North Carolina, Charlotte, an M.S. in geology from Emory University, and a Ph.D. in science education from the University of Georgia.

Eric M. Riggs is the assistant dean for Diversity and Graduate Student Development and a research associate professor of geoscience education at Texas A&M University. Previously, he was the founding co-director of an interdisciplinary research center dedicated to the advancement of science and mathematics education research at Purdue University. His research interests lie at the intersection of cognitive science and geology, especially across diverse cultures and backgrounds. He is especially interested in workforce issues, including Hispanic and Native American diversity in the geosciences. Dr. Riggs is a past President of NAGT. He received a B.A. in English literature from Pomona College and a Ph.D. in geological sciences from the University of California, Riverside.

Karl K. Turekian (*deceased*) was Sterling Professor of Geology and Geophysics Emeritus and a senior research scientist at Yale University. His research focused on the use of radioactive and radiogenic nuclides for deciphering the environmental history of Earth. He was also interested in science education, and previously participated in studies related to this topic. Dr. Turekian served on more than 10 NRC committees and boards. He was a fellow of the American Academy of Arts and Sciences, AGU, AAAS, and the Geological Society of America (GSA), and a member of the National Academy of Sciences. He received an undergraduate degree in chemistry from Wheaton College (Illinois) and a Ph.D. in geochemistry from Columbia University.

Lisa D. White is the director of education and outreach at the University of California Museum of Paleontology. Previously, she was an associate dean in the College of Science and Engineering and a professor of geology in the Geosciences Department at San Francisco State University. Her background is in micropaleontology, paleoceanography, and stratigraphy. Dr. White has extensive experience with science outreach programs for urban students and is active in efforts to increase diversity in the geosciences. She is principal investigator of the SF-ROCKS (Reaching Out to Communities and Kids with Science in San Francisco) program, which is aimed at attracting minority high school students to the geosciences through environmental research projects and training. She coordinated the Minority Participation in the Earth Sciences Program at the U.S. Geological Survey from 1988 to 1995, and is a past chair of the GSA Committee on Minorities and Women in the Geosciences. Dr. White received her Ph.D. in earth sciences from the University of California, Santa Cruz.

Appendix F

Acronyms and Abbreviations

AAAS	American Association for the Advancement of Science
AFRI	Agriculture and Food Research Initiative
AGU	American Geophysical Union
CCI	Community College Internships
DOE	Department of Energy
DOI	Department of the Interior
EdMap	National Cooperative Geologic Mapping Program
EPA	Environmental Protection Agency
EPP	Educational Partnership Program
ESSEA	Earth System Science Education Alliance
FY	fiscal year
GeoEd	Geoscience Education
GEO-Teach	Geoscience Teacher Training
GLOBE	Global Learning and Observations to Benefit the Environment
GRACE	Gravity Recovery and Climate Experiment
GRO	Greater Research Opportunities
GSA	Geological Society of America
HBCUs	historically Black colleges and universities
ISS	International Space Station
LASER	Leadership and Assistance for Science Education Reform

MSI	minority-serving institution
NAGT	National Association of Geoscience Teachers
NASA	National Aeronautics and Space Administration
NOAA	National Oceanic and Atmospheric Administration
NPS	National Park Service
NRC	National Research Council
NSF	National Science Foundation
OEDG	Opportunities for Enhancing Diversity in the Geosciences
REU	Research Experience for Undergraduates
SAGE	Summer of Applied Geophysical Experience
SCGF	Office of Science Graduate Fellowship
STAR	Science to Achieve Results
STEM	science, technology, engineering, and mathematics
SULI	Science Undergraduate Laboratory Internships
TESSE	Transforming Earth Systems Science Education
USDA	U.S. Department of Agriculture
USGS	U.S. Geological Survey